Automated Deep Learning Using Neural Network Intelligence

Develop and Design PyTorch and TensorFlow Models Using Python

Ivan Gridin

Apress®

Automated Deep Learning Using Neural Network Intelligence: Develop and Design PyTorch and TensorFlow Models Using Python

Ivan Gridin
Vilnius, Lithuania

ISBN-13 (pbk): 978-1-4842-8148-2
ISBN-13 (electronic): 978-1-4842-8149-9
https://doi.org/10.1007/978-1-4842-8149-9

Managing Director, Apress Media LLC: Welmoed Spahr
Acquisitions Editor: Celestin Suresh John
Development Editor: Laura Berendson
Coordinating Editor: Mark Powers

Cover designed by eStudioCalamar

Cover image by Vecteezy (www.vecteezy.com)

Distributed to the book trade worldwide by Apress Media, LLC, 1 New York Plaza, New York, NY 10004, U.S.A. Phone 1-800-SPRINGER, fax (201) 348-4505, e-mail orders-ny@springer-sbm.com, or visit www.springeronline.com. Apress Media, LLC is a California LLC and the sole member (owner) is Springer Science + Business Media Finance Inc (SSBM Finance Inc). SSBM Finance Inc is a **Delaware** corporation.

For information on translations, please e-mail booktranslations@springernature.com; for reprint, paperback, or audio rights, please e-mail bookpermissions@springernature.com.

Apress titles may be purchased in bulk for academic, corporate, or promotional use. eBook versions and licenses are also available for most titles. For more information, reference our Print and eBook Bulk Sales web page at http://www.apress.com/bulk-sales.

Any source code or other supplementary material referenced by the author in this book is available to readers on GitHub (https://github.com/Apress). For more detailed information, please visit http://www.apress.com/source-code.

Printed on acid-free paper

To Ksenia and Elena

Table of Contents

About the Author

Ivan Gridin is a researcher, author, developer, and artificial intelligence expert who has worked on distributive high-load systems and implemented different machine learning approaches in practice. One of the primary areas of his research is the design and development of predictive time series models. Ivan has fundamental math skills in random process theory, time series analysis, machine learning, reinforcement learning, neural architecture search, and optimization. He has published books on genetic algorithms and time series forecasting.

He is a loving husband and father and collector of old math books.

You can learn more about him on LinkedIn: `https://www.linkedin.com/in/survex/`.

About the Technical Reviewer

Andre Ye is a deep learning researcher and writer working toward making deep learning more accessible, understandable, and responsible through technical communication. He is also a cofounder at Critiq, a machine learning platform facilitating greater efficiency in the peer-review process. In his spare time, Andre enjoys keeping up with current deep learning research, reading up on history and philosophy, and playing the piano.

Introduction

Machine learning is a big part of our lives in today's world. We cannot even think of a world without machine learning approaches at the moment, and it has already started to take a huge part of our daily activities. Websites, mobile applications, self-driving cars, home devices, and many things surrounding us use machine learning algorithms. The dawn of computing power, especially the graphics processing unit, was accompanied by the practical start of the deep learning implementation. Deep learning studies the design of deep neural networks. This approach shows impressive efficiency and has experienced explosive growth in recent years.

Not surprisingly, the number of tasks to be solved and the need for machine learning specialists are constantly growing. At the same time, the number of routine actions that developers and data scientists execute to solve machine learning problems is increasing. Meanwhile, researchers developed special techniques to save time and automate the most common machine learning tasks. These techniques were separated into the special area called automated machine learning, or AutoML. This book focuses on the automated deep learning (AutoDL) area, which studies the automation of deep learning problems. AutoDL considers the issues of creating and designing optimal deep learning models. This approach has been rapidly developed in recent years and, in some cases, can completely automate the solution of typical tasks.

This book is about implementing AutoDL methods using Microsoft Neural Network Intelligence (NNI). NNI is a Python toolkit that contains the most common and advanced AutoDL methods: Hyperparameter Optimization (HPO), Neural Architecture Search (NAS), and Model Compression. NNI supports the most popular deep learning frameworks. This book covers the NNI implementation of various AutoDL techniques using the PyTorch and TensorFlow frameworks.

Chapter 1 focuses on automated deep learning basics and why we should put this approach into practice. We will also install NNI and examine the main basic scenarios for its use. We will learn how to run simple Hello World Experiments and interact with NNI via the command line and WebUI.

In Chapter 2, we will move on to the study of the most common AutoDL task – Hyperparameter Optimization (HPO). We will learn what Hyperparameter Optimization is, what hyperparameters are, and how to organize an NNI HPO experiment using PyTorch and TensorFlow. We will also construct three kinds of research that will make a historical journey to the origins of deep learning. The first one will help us determine the best LeNet model hyperparameters for the MNIST problem. The second research integrates a new dropout layer and rectified linear unit (ReLU) activation into the original LeNet model. And the third one will show us how we can evolve the LeNet model in AlexNet using simple HPO techniques.

In Chapter 3, we will study NNI's main search algorithms (Tuners), which aim to solve HPO tasks. Here, we will consider the practical application and the description of the following algorithms: Evolution Tuner, Anneal Tuner, and SMBO Tuners. Chapter 3 provides the creation of a custom Tuner and applies it to the classic Shallow AutoML problem – building an optimal pipeline using Scikit methods.

In Chapter 4, we will begin to research Neural Architecture Search (NAS). NAS is an approach that studies the creation and design of neural networks best suited to solve a specific problem. This chapter covers Multi-trial NAS and its main principles. We'll discuss the NNI Retiari framework, define Model Spaces and Model Mutators, and set up experiments that construct optimal neural networks. Also, this chapter introduces various exploration algorithms that explore Multi-trial NAS Model Space: Regularized Evolution, TPE Strategy, and RL Strategy. Next, we will build LeNet-based and ResNet-based Multi-trial NAS experiments to solve the CIFAR-10 problem.

In Chapter 5, we move on to One-shot NAS, one of the latest advances in AutoDL. This chapter explains how to construct a Supernet, how to design cell-based neural architectures, and perform Efficient Neural Architecture Search (ENAS) and Differentiable Architecture Search (DARTS) One-shot NAS algorithms.

In Chapter 6, we will cover the important topic of model pruning. Model pruning compresses neural network removing redundant weights or even layers. This technique is crucial for lightweight devices when we need to save computing resources. This chapter will examine basic One-shot and iterative pruning algorithms.

Chapter 7 will focus on practical recipes for using NNI to organize robust, extensive, and big data experiments.

This book explores practical NNI applications of AutoDL methods and describes their theory also. Therefore, this book can be helpful for data scientists who want to get the idea that underlies various AutoDL techniques and algorithms.

This book requires intermediate deep learning understanding and TensorFlow or PyTorch knowledge.

Source Code Listings

This book has many practical examples and code listings. Source code listings accompany each chapter in this book. You can download the source code from the following GitHub repository: `https://github.com/Apress/automated-deep-learning-using-neural-network-intelligence`.

CHAPTER 1

Introduction to Neural Network Intelligence

There was a great burst of deep learning industry in the past few years. Deep learning approaches have achieved outstanding results in computer vision, natural language processing, robotics, time series forecasting, and optimal control theory. However, there is no "silver bullet model" to solve all kinds of problems. Each problem and dataset needs a specific model architecture to achieve suitable performance. Machine learning models, especially deep learning models, have a lot of tunable parameters that can drastically affect the model performance. Those are model design, training method, model configuration hyperparameters, etc. The model optimization process is performed for each application and even each dataset. Data scientists and machine learning experts often spend a lot of time performing manual model optimization. This activity can be frustrating because it takes too much time and is usually based on an expert's experience and quasi-random search.

However, recent results in automated machine learning and deep learning meta-optimization make it possible to automate the optimizing process for a specific task. It is also possible to create brand new model architecture from scratch without having any experience solving similar problems in the past. The Neural Network Intelligence (**NNI**) toolkit provides the latest state-of-the-art techniques to solve the most challenging automated deep learning problems. We'll start exploring the basic NNI features in this chapter.

What Is Automated Deep Learning?

Before we dive into NNI techniques, let's talk about automated deep learning, examine its use cases, and why you need it. Modern machine learning models can contain enormous complexity in their design. Architecture can have thousands of adjustable

I. Gridin, *Automated Deep Learning Using Neural Network Intelligence*,
https://doi.org/10.1007/978-1-4842-8149-9_1

parameters and connections between different neural layers. It is computationally impossible to test each architecture hyperparameter combination to select the best one. However, modern graphics processing units can already significantly speed up the training of machine and especially deep learning models, which means that many machine learning processes can be automated. Therefore, a new domain in machine learning has appeared called automated machine learning (**AutoML**). AutoML deals with tasks that automate optimal machine learning model production. The area of AutoML is very young and is growing rapidly. Machine learning can be divided into shallow learning and deep learning. Shallow learning contains classical methods: random forest, support vector machine, k-nearest neighbors, etc. In comparison, deep learning studies the construction of neural networks based on convolution layers, linear layers, pooling, splitting and joining connections, etc. Shallow learning and deep learning contain similar automated machine learning techniques, but their application differs significantly. Therefore, we can highlight a separate area of AutoML, which deals only with deep learning – this area is called automated deep learning (**AutoDL**). There are four main sections for automated deep learning:

- Hyperparameter Optimization (**HPO**)
- Neural Architecture Search (**NAS**)
- Feature Engineering
- Model Compression

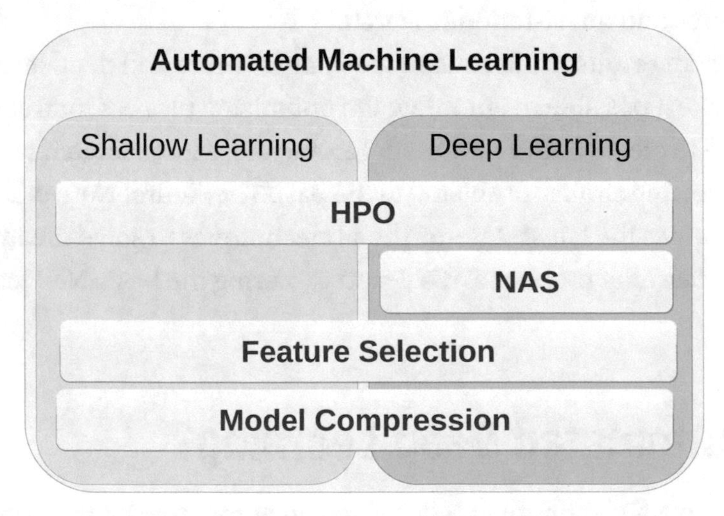

Figure 1-1. *AutoDL sections*

Let's move on and discuss what exactly we need AutoDL for.

No Free Lunch Theorem

And I want to start with the following fundamental statement, called the No Free Lunch (**NFL**) theorem. No Free Lunch theorem states that

> *Any two optimization algorithms are equivalent*
> *when their performance is averaged across all possible problems.*
>
> —David Wolpert

Let's elaborate NFL theorem to more functional language. Say we have a set of datasets: $\boldsymbol{D_1}, \boldsymbol{D_2}, \boldsymbol{D_3}, \ldots$, and the estimated performance of random search algorithm $\boldsymbol{R}$ on each dataset $\boldsymbol{D_i}$ equals to r:

$$E(\boldsymbol{R}; \boldsymbol{D}_i) = r, \text{ for any } I \tag{1}$$

Then for any search algorithm $\boldsymbol{A}$ and any dataset $\boldsymbol{D_i}$ with estimation $r + q$, there is a dataset $\boldsymbol{D_j}$ with estimation $r - q$:

$$E(\boldsymbol{A}; \boldsymbol{D}_i) = r + q, \; E(\boldsymbol{A}; \boldsymbol{D}_j) = r - q \tag{2}$$

Statement 2 says that if algorithm $\boldsymbol{A}$ is better than random algorithm $\boldsymbol{R}$ for dataset $\boldsymbol{D_i}$, then there is dataset $\boldsymbol{D_j}$ for which algorithm $\boldsymbol{A}$ will be worse than random algorithm $\boldsymbol{R}$ and $E(\boldsymbol{A}; \boldsymbol{D_i}) + E(\boldsymbol{A}; \boldsymbol{D_j}) = E(\boldsymbol{R}; \boldsymbol{D_i}) + E(\boldsymbol{R}; \boldsymbol{D_j})$. This fact makes all algorithms equivalent if we consider them separately from a specific dataset and task. For example, let's say we have an algorithm $\boldsymbol{A}$ for predicting the color of the next box by previous ones with rules listed in Table 1-1.

Table 1-1. *Box prediction algorithm*

Rule	Previous Box	Current Box	Prediction
1	Black	Black	White
2	Black	White	White
3	White	Black	Black
4	White	White	Black

And the prediction algorithm **A** works with 100% accuracy for dataset $\mathbf{D}_1$, in which two black boxes follow two white boxes and two white boxes follow two black boxes, as shown in Figure 1-2.

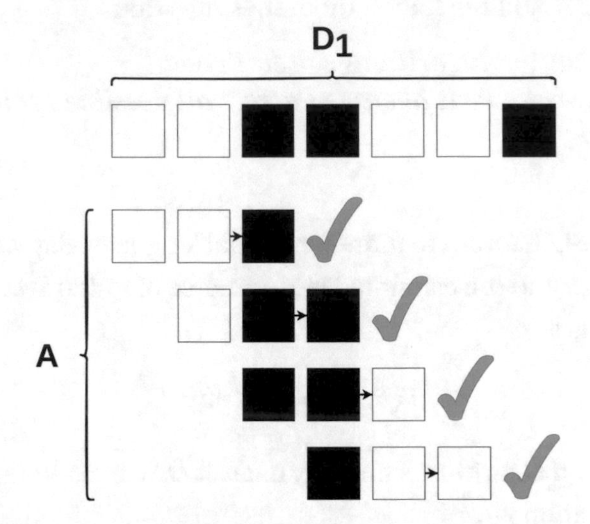

Figure 1-2. *Performance of prediction algorithm **A** on dataset $\mathbf{D}_1$: 100% accuracy*

But let's examine how algorithm **A** works on dataset $\mathbf{D}_2$, in which white and black boxes alternate one after another one by one, as shown in Figure 1-3.

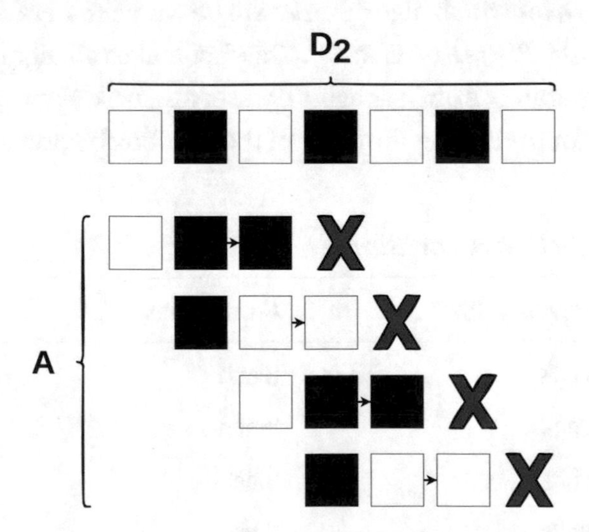

Figure 1-3. *Performance of prediction algorithm **A** on dataset $\mathbf{D}_2$: 0% accuracy*

Figure 1-3 demonstrates that algorithm **A** has 0% accuracy on dataset $\mathbf{D_2}$. This example illustrates that "there is no optimal algorithm for all datasets" and "there is no optimal solution for all cases." And how does NFL theorem influence deep learning? Each deep learning model and each dataset generates a loss function that should be minimized. If we have two deep learning models, $\mathbf{M_1}$ and $\mathbf{M_2}$, then this means that they show good results only for certain types of problems and certain types of datasets. You cannot expect the same deep learning model to perform similarly on a different dataset, much less for a different kind of problem. So if you apply model $\mathbf{M_1}$ and model $\mathbf{M_2}$ to problem $\mathbf{P_1}$ on dataset $\mathbf{D_1}$, you can expect that model $\mathbf{M_1}$ will show good performance in this case, and model $\mathbf{M_2}$ will demonstrate poor performance. Figure 1-4 illustrates this point.

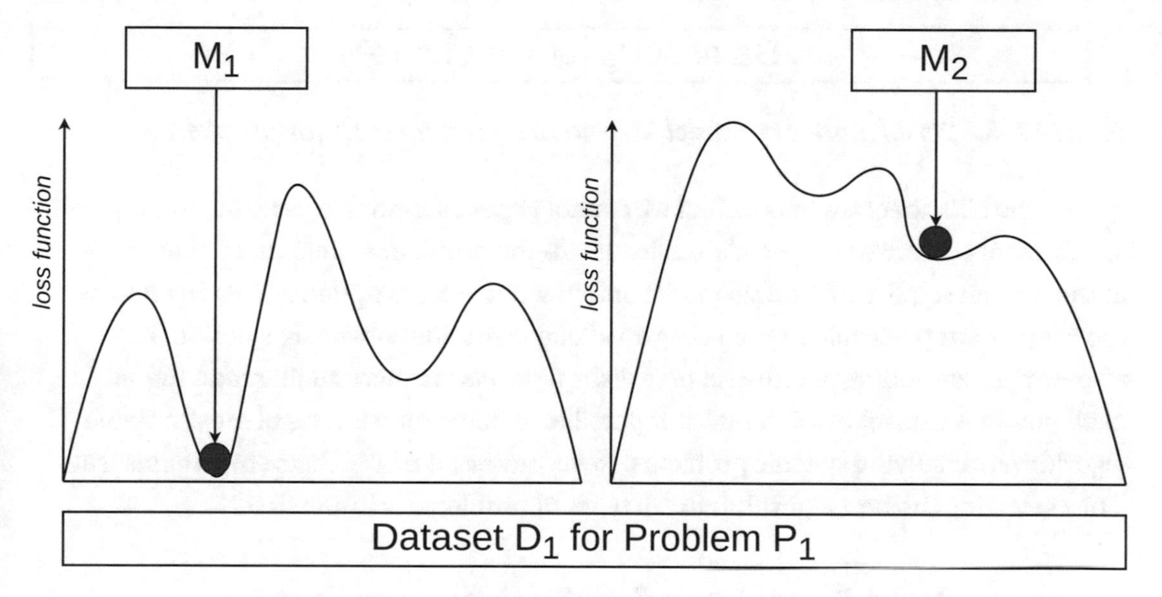

***Figure 1-4.** Performance of model M_1 and M_2 on dataset D_1 for problem P_1*

But we can get the opposite results if the models are applied to a different problem and a different dataset as shown in Figure 1-5.

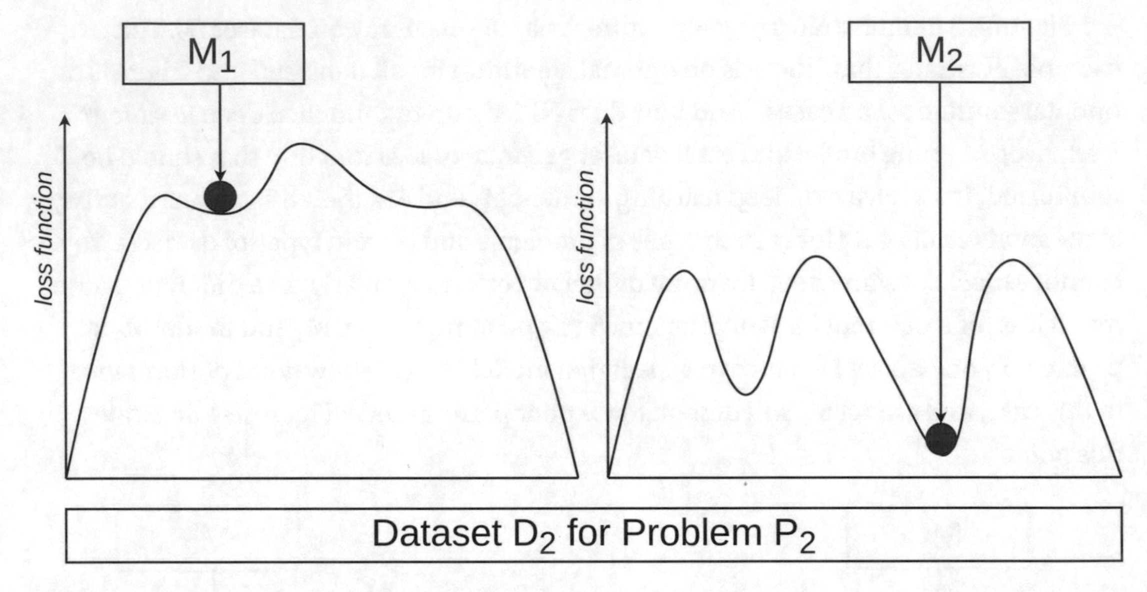

***Figure 1-5.** Performance of model M_1 and M_2 on dataset D_2 for problem P_2*

So, the NFL theorem tells us that we cannot expect a model to perform equally well for different cases. The slightest modification in the problem statement or changes in the dataset require additional model optimization for the updates. This fact makes the AutoDL irreplaceable in preparing an effective production-ready solution. It is also worth mentioning that the set of realistic datasets is much smaller than the set of all possible datasets, which makes it possible to determine a class of most suitable algorithms for solving specific problems. Nevertheless, the NFL theorem remains true since selecting the best algorithm for all types of problems is impossible.

Injecting New Deep Learning Techniques into Existing Model

Suppose we have a deep learning model that performs well and shows satisfactory results. Later, a new deep learning technique appeared that could significantly improve the performance of our model. It could be a special deep learning layer, block, cell, or a new activation function. But we do not know how exactly to inject that technique into the model architecture. This can be accomplished with AutoDL, which will make optimal use of the new technique in the current deep learning model design. Figure 1-6 illustrates this approach.

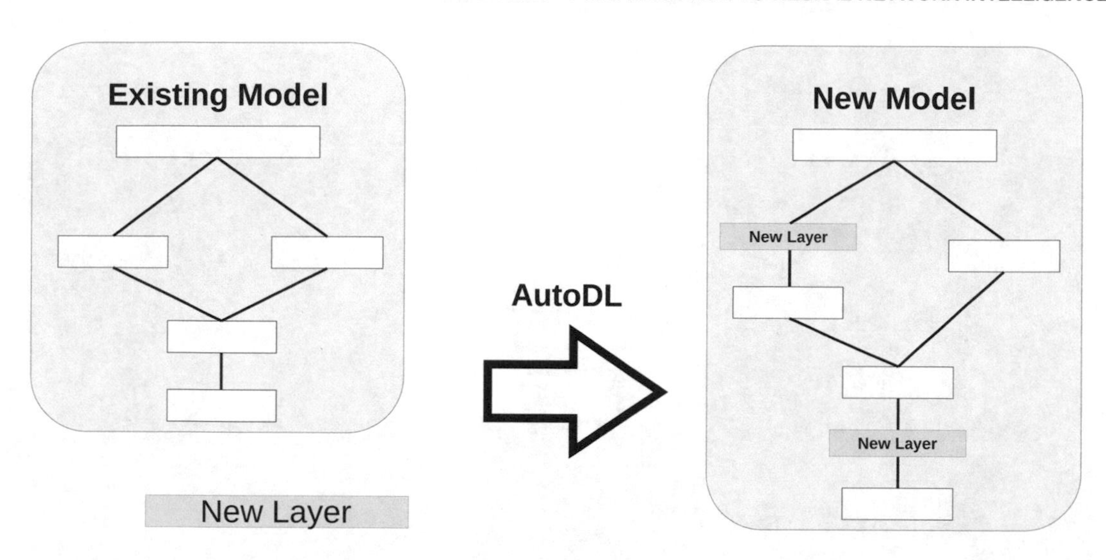

Figure 1-6. *Injecting new deep learning technique*

This approach will help update the model with the latest advances in deep learning, enhancing the model's performance.

Adjusting Model to a New Dataset

Let's say we have a model that solves the problem of energy consumption prediction in New York. The model has been trained on the historical dataset and works well. We have decided to port this model for energy consumption prediction in Berlin. We expect this model to perform for Berlin as well as it performed for New York. But people in another country may have a little bit different habits and behaviors that affect the original model's ability to capture patterns correctly. Therefore, it would be good to customize the original model for the new Berlin historical dataset. Figure 1-7 demonstrates how the existing model could be adapted to a new dataset, updating some of its hyperparameters like convolution layer filters, linear layer features, etc.

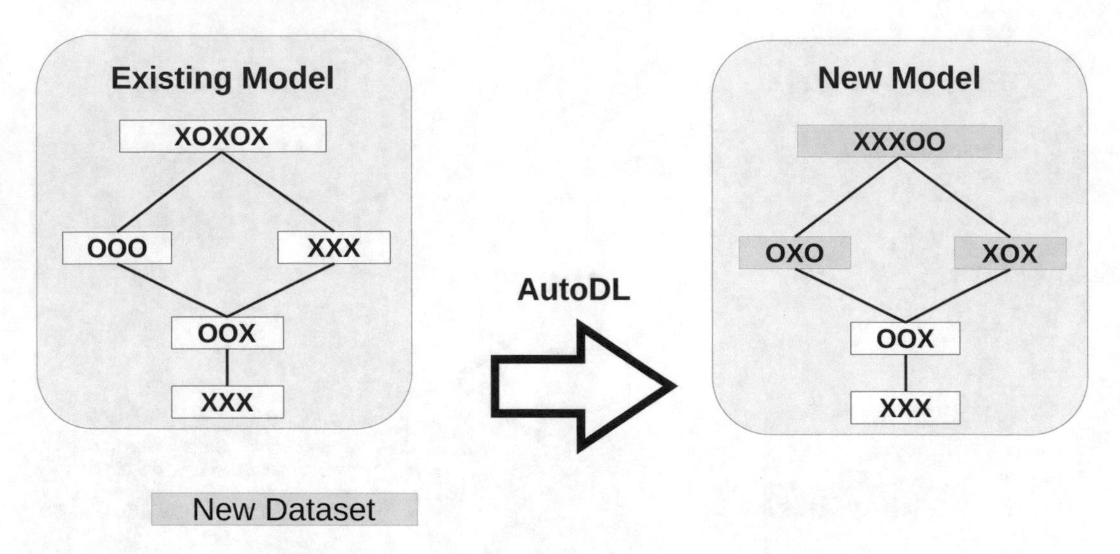

Figure 1-7. *Adapting model to new dataset*

Using AutoDL techniques, you can adapt the model to other datasets.

Creating a New Model from Scratch

And this is the most exciting part. Let's say we have a task, and there is no idea about the architecture of a neural network that could cope with it. We can borrow some ideas from other tasks, do manual investigations, study the statistical properties of the dataset, etc. But at the moment, there are Neural Architecture Search (**NAS**) methods that allow you to build a production-ready neural network from scratch, as shown in Figure 1-8.

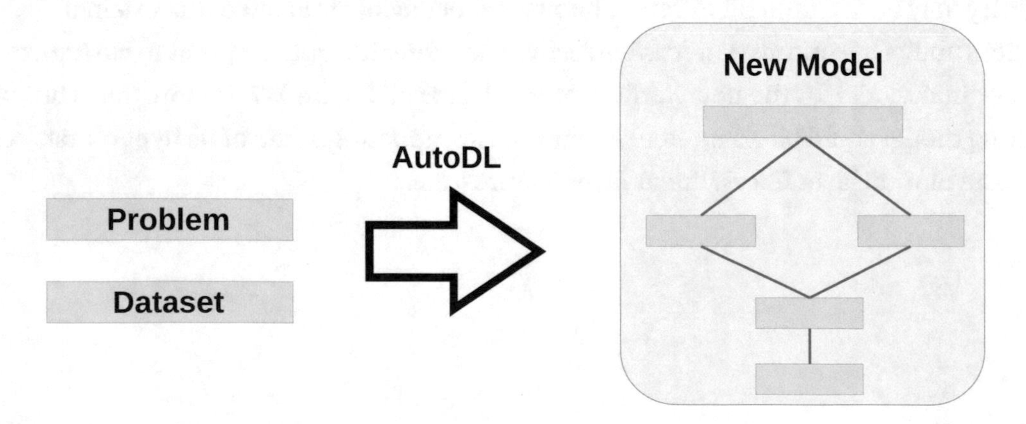

Figure 1-8. *Neural Architecture Search*

And I find this to be a fantastic direction for further research and practical applications. Humanity has developed deep learning models and their training based on error backpropagation. Neural networks of a particular architecture can reveal the most complex dependencies and patterns. So why not take the next step and develop neural network design algorithms that will create the optimal neural network architecture for a specific task.

Reinventing the Wheel

Many machine learning experts spend a lot of time developing existing methods to solve the problems described earlier. Automated machine learning techniques can save weeks or even months of development. Of course, automated deep learning cannot substitute for deep learning engineers, and human experience and intuition is the main driver in all inventions nowadays. But anyway, AutoDL can significantly decrease the amount of custom work needed. Automated deep learning should become a must-have tool for solving practical problems that can save significant time.

Working with Source Code

This book demonstrates many practical examples, accompanied by source code that can be downloaded from the following GitHub repository. This book has many practical examples and code listings. The chapter's source code listings accompany each chapter in this book. You can download the source code from the following GitHub repository: **`https://github.com/Apress/automated-deep-learning-using-neural-network-intelligence`**. Most of the listings in the book are presented in source code. All commands in the book are run relative to the root of the source code folder.

Neural Network Intelligence Installation

Neural Network Intelligence (**NNI**) is a powerful toolkit to help users solve AutoML. NNI manages search processes, visualizes results, and distributes AutoML jobs to different machine learning platforms.

Install

NNI minimal system requirements are: Ubuntu, 18.04; macOS, 11; Windows 10, 21H2 and Python 3.7.

NNI can be simply installed as follows:

```
pip install nni
```

We will be using version 2.7 in this book, so I highly recommend installing version 2.7 to avoid version differences:

```
pip install nni==2.7
```

Let's test the installation by executing "*Hello World*" scenario. Run the following command (`ch1/install/hello_world/config.yml` file is contained in the source code):

```
nnictl create --config ch1/install/hello_world/config.yml
```

If the installation was successful, you should see the following output:

```
INFO:  Starting restful server...
INFO:  Successfully started Restful server!
INFO:  Starting experiment...
INFO:  Successfully started experiment!
The experiment id is <EXPERIMENT_ID>
The Web UI urls are: http://127.0.0.1:8080
```

And you can follow the link `http://127.0.0.1:8080` in your browser. Figure 1-9 demonstrates NNI web user interface that we will cover in the next sections.

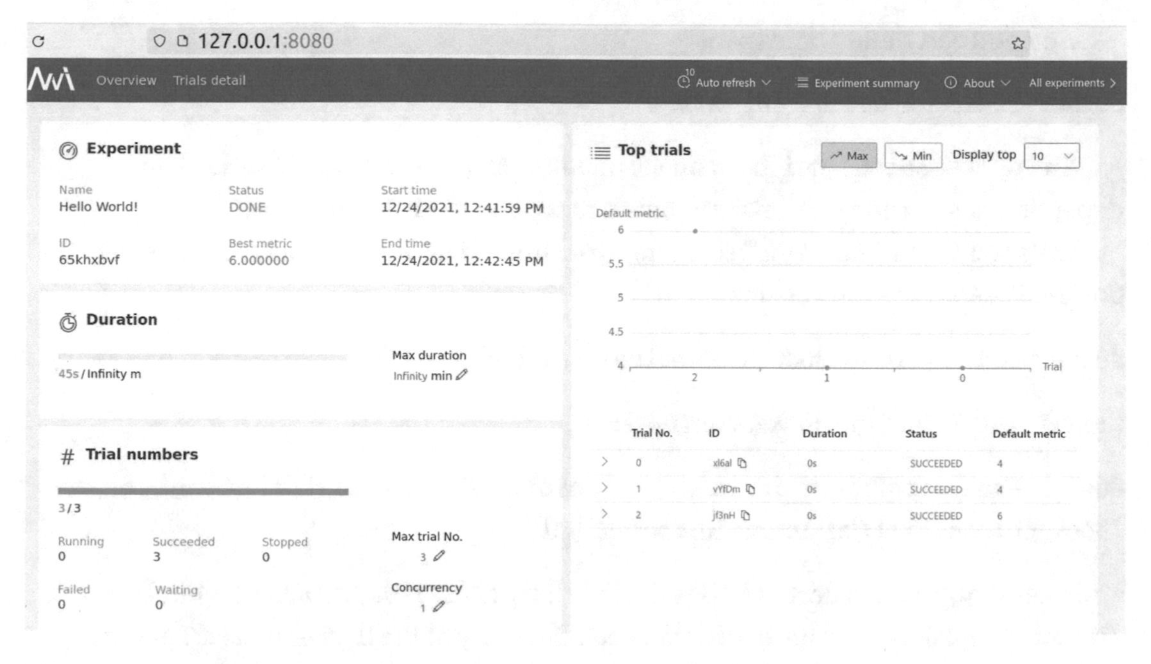

Figure 1-9. *NNI WebUI*

If everything is ok, then you can stop NNI by executing the following in the command line:

```
nnictl stop
```

Docker

If you have any problems with the installation, you can use the docker image that was prepared for this book. The `Dockerfile` in Listing 1-1 is based on the official NNI docker image `msranni/nni:v2.7` from the official docker repository: `https://hub.docker.com/r/msranni/nni/tags`.

Listing 1-1. NNI Dockerfile for the book.

```
FROM msranni/nni:v2.7
RUN mkdir /book
ADD . /book
EXPOSE 8080
ENTRYPOINT ["tail", "-f", "/dev/null"]
```

And you can build an image as

```
docker build -t autodl_nni_book .
```

The docker autodl_nni_book image contains all the necessary libraries and dependencies to run all the experiments that we will study in this book.

Let's run the "*Hello World*" scenario we examined in the previous section using docker. We start the docker container:

```
docker run -t -d -p 8080:8080 autodl_nni_book
```

then we run NNI in the docker container:

```
docker exec <container_id> bash -c "nnictl create --config
/book/ch1/install/hello_world/config.yml"
```

and after that, you can access NNI WebUI via http://127.0.0.1:8080 in your browser. The code repository for this book is in /book directory of the docker image. Therefore, in the autodl_nni_book docker image, you can execute all commands that will concern NNI as follows:

```
docker exec <container_id> bash -c "nnictl <nni_command>"
```

But in any case, the docker's capabilities are limited. For flexible debugging and better interaction with NNI, I strongly recommend that you work with NNI without using the docker if possible.

Search Space, Tuner, and Trial

Let's take a quick look at one core NNI concept. When we optimize a model, we select a particular set of parameters that determine the operation of our model. **Search space** defines this set of parameters. Search space is a key concept in automatic machine learning. The search space contains all possible parameters and architectures that are hypothetically acceptable for the optimized model.

Although the search space contains a finite number of parameters, nevertheless in most cases, it is practically impossible to test all parameters from the search space. The search space is too large. Therefore, a special component called **Tuner** is applied in selecting the most appropriate and promising parameters for testing. Tuner estimates the results and selects new parameters to check their suitability for model optimization.

Tuner selects a parameter in the search space and transfers it to **Trial**. Trial is a Python script that tests the model with parameters passed by Tuner and returns a metric that estimates the model's performance.

This search process can be depicted as shown in Figure 1-10.

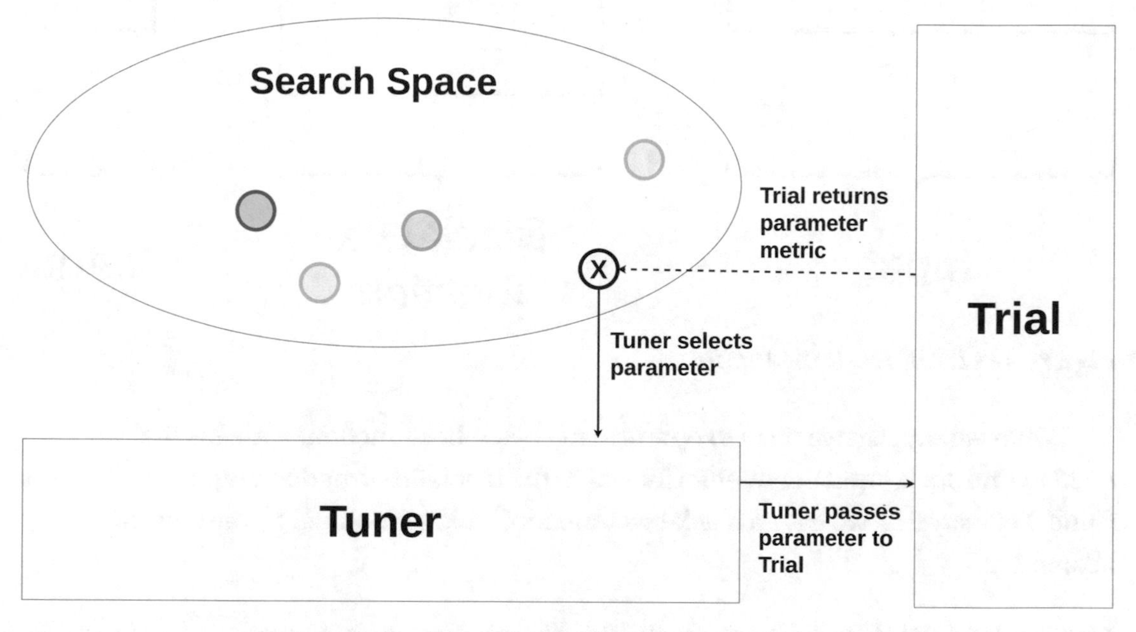

Figure 1-10. *Search space, Tuner, and Trial*

After a certain number of trials, we have a sufficient number of results that estimate the suitability of each parameter for an optimized model.

Black-Box Function Optimization

Let's examine how NNI works by optimizing a black-box function. A black-box function is a function that takes input parameters and returns a value, but we have no idea what is going on under the function's hood. Sometimes, we know how a black-box function acts and, in some cases, even know its formula. But the nature of this function is so complicated that the analytical study is too challenging.

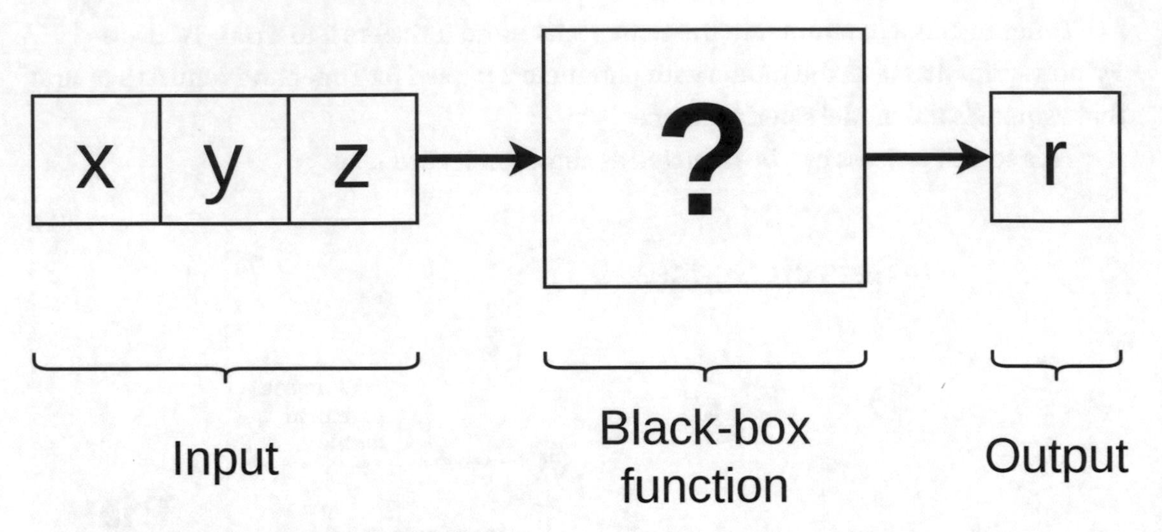

Figure 1-11. *Black-box function*

When we say that we need to optimize the black-box function, it means that we need to find such input parameters for which the black-box function outputs the highest value. Let's say that we have a black-box function, which is defined by the code in Listing 1-2.

Listing 1-2. Black-box function. ch1/bbf/black_box_function.py

```
from math import sin, cos
def black_box_function(x, y, z):
    """
    x in [1, 100] integer
    y in [1, 10] integer
    z in [1, 10000] real
    """
    if y % 2 == 0:
        if x > 50:
            r = (pow(x, sin(z)) - x) * x / 2
        else:
            r = (pow(x, cos(z)) + x) * x
    else:
        r = pow(y, 2 - sin(x) * cos(z))
    return round(r / 100, 2)
```

Of course, the optimization problem for the function presented in Listing 1-2 can be solved analytically, but let's pretend that we do not know how the function acts inside the black box. All we know is that the black-box function returns real value and receives the following input parameters:

- x positive integer from 1 to 100
- y positive integer from 1 to 10
- z float from 1 to 10 000

Let's start solving our problem by defining a search space. Search space is defined in JSON format using special directives. We will define the search space using the following JSON file.

Listing 1-3. Search space. ch1/bbf/search_space.json

```
{
  "x": {"_type": "quniform", "_value": [1, 100, 1]},
  "y": {"_type": "quniform", "_value": [1, 10, 1]},
  "z": {"_type": "quniform", "_value": [1, 10000, 0.01]}
}
```

`quniform` directive creates a value list from `a` to `b` with step `s`. So the search space defined in Listing 1-3 can be presented the following way:

- x in [1, 2, 3, ..., 99, 100]
- y in [1, 2, 3, ..., 9, 10]
- z in [1, 1.01, 1.02, ..., 9 999.99, 10 000]

Note We'll explore how to define search space in more detail in the next chapter.

Now let's move on to the trial definition.

Listing 1-4. Trial. ch1/bbf/trial.py

```
import os
import sys
import nni
```

```
# For NNI use relative import for user-defined modules
SCRIPT_DIR = os.path.dirname(os.path.abspath(__file__)) + '/../..'
sys.path.append(SCRIPT_DIR)

from ch1.bbf.black_box_function import black_box_function

if __name__ == '__main__':
    # parameter from the search space selected by tuner
    p = nni.get_next_parameter()
    x, y, z = p['x'], p['y'], p['z']
    r = black_box_function(x, y, z)
    # returning result to NNI
    nni.report_final_result(r)
```

Trial receives parameter from NNI using `nni.get_next_parameter` function and returns the metric using `nni.report_final_result`. Trial takes NNI parameters, passes them to the black-box function, and returns the result. Listing 1-4 has to be run by the NNI server, so you will get an error if you try to run it.

Note We'll explore how to define trial in more detail in the next chapter.

And the last thing left for us to do is to define the configuration of our experiment, which will look for the best input parameters for the black-box function.

Listing 1-5. Experiment configuration. ch1/bbf/config.yml

```
trialConcurrency: 4
maxTrialNumber: 1000
searchSpaceFile: search_space.json
trialCodeDirectory: .
trialCommand: python3 trial.py
tuner:
  name: Evolution
  classArgs:
      optimize_mode: maximize
trainingService:
  platform: local
```

The experiment that we have defined in Listing 1-5 has the following properties:

- Four-thread pool for trial execution.
- The maximum number of trials is 1000.
- Search space is defined in `search_space.json`.
- Trial is executed by running `python3 trial.py`.
- NNI will use a Tuner based on genetic algorithms.

Note We'll explore how to define experiment configuration in more detail in the next chapter.

Now everything is ready to find the input parameters that maximize the black-box function. Let's run NNI:

```
nnictl create --config ch1/bbf/config.yml
```

And you can monitor the experiment process in the web panel: `http://127.0.0.1:8080`.

Note Typically, testing a deep learning model architecture takes about a few minutes, and the NNI is optimized for longer trials. Therefore, the NNI is not well suited for high-speed tests, and executing the value of the black-box function can take more time than expected. This is due to the data exchange mechanism between the main NNI process and its sub-processes. If you want to shorten the experiment execution time, change the `maxTrialNumber` parameter to `100` in ch1/bbf/config.yml.

After completing the experiment, you can observe the parameter that returned the best metric on the NNI overview page: `http://127.0.0.1:8080/oview`.

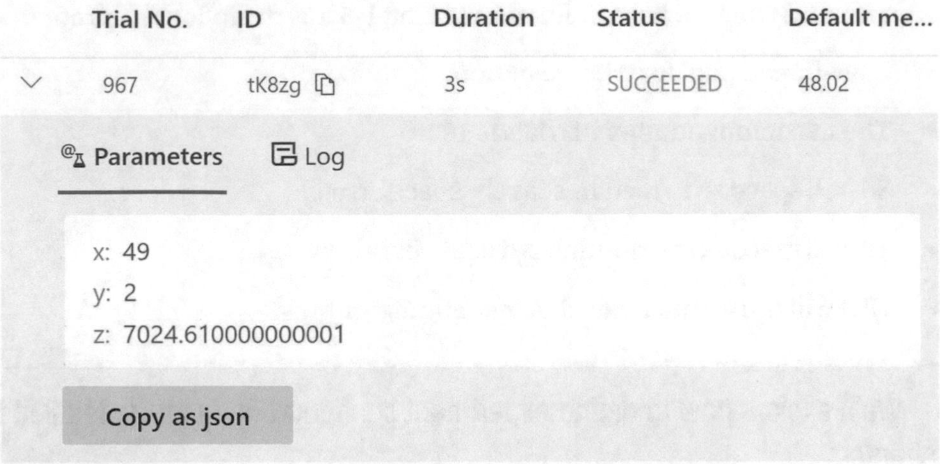

	Trial No.	ID	Duration	Status	Default me...
∨	967	tK8zg	3s	SUCCEEDED	48.02

Parameters Log

x: 49
y: 2
z: 7024.610000000001

Copy as json

***Figure 1-12.** NNI best trial for black-box function optimization*

We see that parameter (`x=49, y=2, z=7024.61`) is the best result of the experiment. The function for this parameter returns `48.02`, which is the maximum of all trials. Of course, we could have obtained the same result more simply, but now, we are introducing the basic capabilities of NNI. In the next chapters, we will see the full strength of this tool.

Web User Interface

Even though NNI allows you to save trial results and later analyze them, NNI provides a convenient web user interface for experiment monitoring and analyzing its results. Let's explore the main features of this web panel.

Overview Page

The overview page `http://127.0.0.1:8080/oview` contains summary information about a running experiment.

The upper left panel contains information about the experiment state (Figure 1-13).

Figure 1-13. *Experiment state panel*

The lower left panel shows the number of trials performed and the running time. The maximum number of trials and the maximum time can be edited on the fly (Figure 1-14).

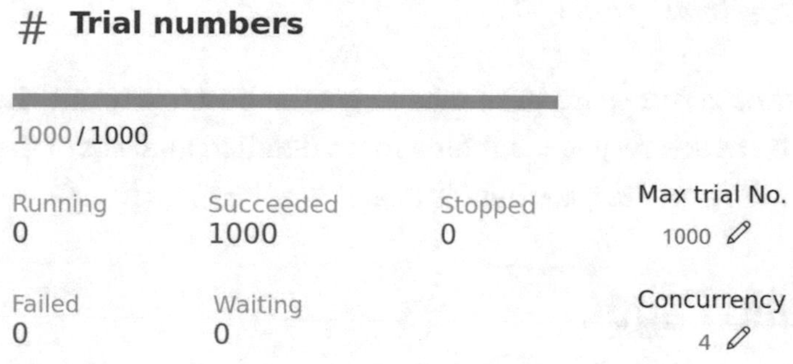

Figure 1-14. *Trial numbers panel*

The right panel on the overview page contains a summary of the top trials (Figure 1-15).

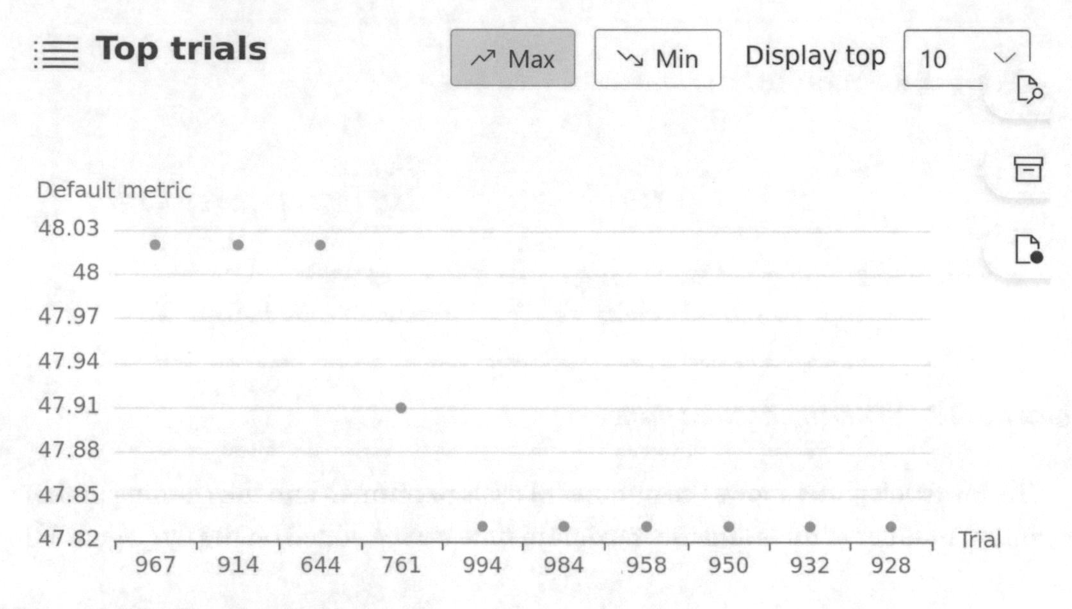

Figure 1-15. Top trials panel

If you just want to run an experiment and get the best test result, then you can only deal with the overview page. But for a more detailed analysis of the experiment execution, you will need the trials details page.

Trials Details Page

The trials details page `http://127.0.0.1:8080/detail` contains a handy visualization of the trials performed. The metric panel contains a history of trials and their metrics. This panel becomes very useful if you toggle the optimization curve. Then you can observe the tuner search progress (Figure 1-16).

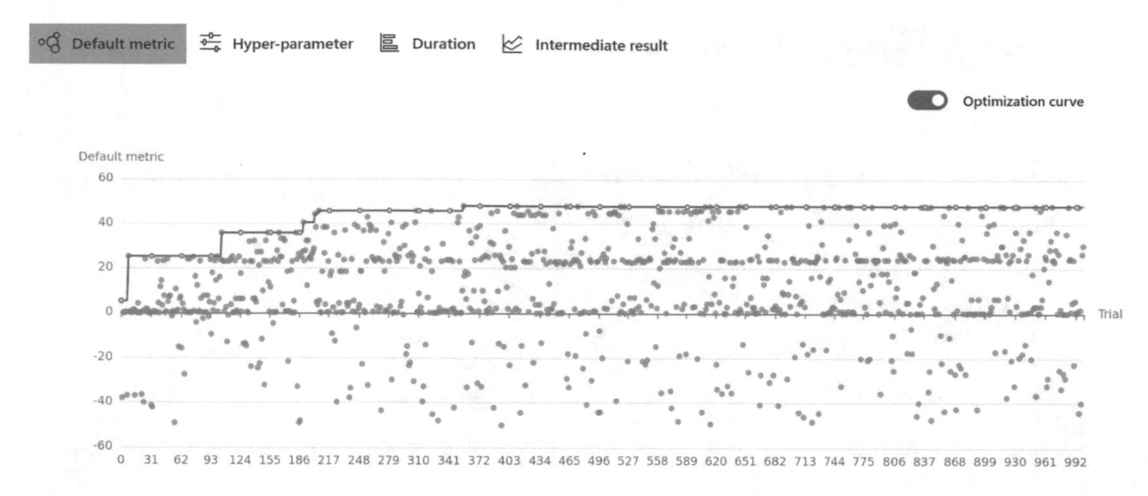

Figure 1-16. *Metric panel*

The hyperparameter panel contains one of the most valuable visualizations. It displays the relationship between the input parameters and the test metric (Figure 1-17).

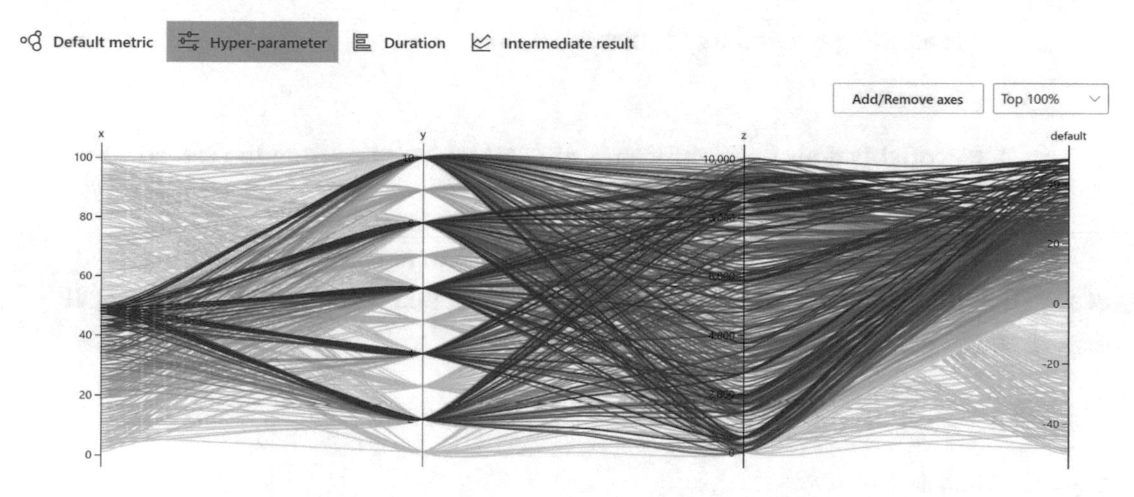

Figure 1-17. *Hyperparameter panel*

This panel allows hyperparameter data mining to help you better understand the nature of the investigated black-box function. We will stay on this moment for a while. Select the top 5% trials on the hyperparameter panel (Figure 1-18).

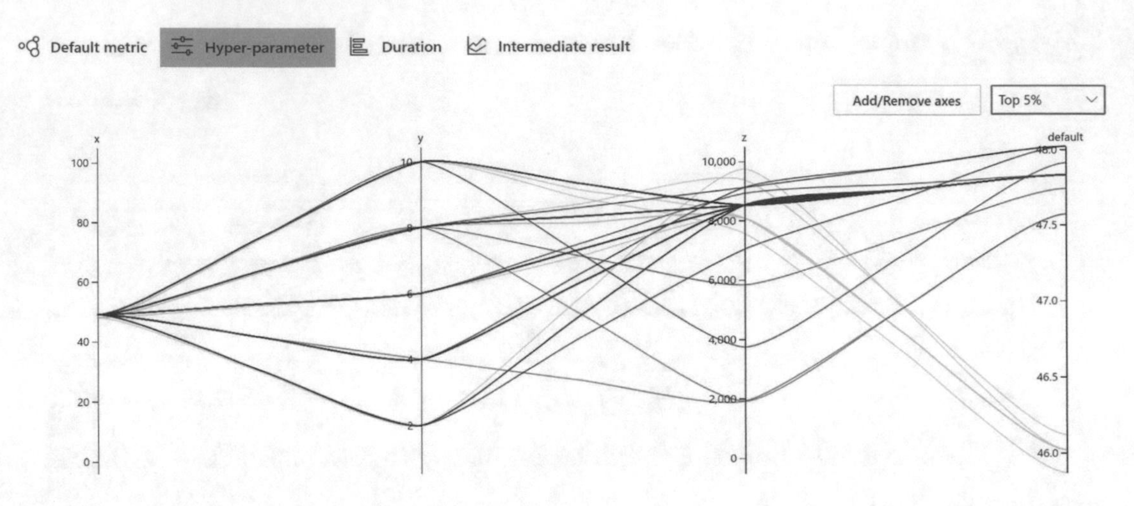

Figure 1-18. Hyper-parameter panel. Top 5%

And we can get a lot of insights from Figure 1-18. For all top 5% trials, the following is true:

- x is an integer less than 51, that is, `50, 49, 48`.
- y is even.
- z probably does not significantly affect the black-box function return value, or further research may be needed.

Based on the information we obtained here, we can perform our own simplified search that finds the best parameter close to `48.02`, which was found during the NNI experiment. Let's examine Listing 1-6.

Listing 1-6. Black-box function optimization. ch1/bbf/custom_search.py

```
import random
from ch1.bbf.black_box_function import black_box_function

seed = 0
random.seed(0)
max_ = -100
best_trial = None
for _ in range(100):
    x = random.choice([50, 49, 48])
    y = random.choice([2, 4, 6, 8, 10])
```

```
    z = round(random.uniform(1, 10_000), 2)
    r = black_box_function(x, y, z)
    if r > max_:
        max_ = r
        best_trial = f'(x={x}, y={y}, z={z}) -> {r}'

print(best_trial)
```

Listing 1-6 returns `(x = 50, y = 2, z = 4756.58) -> 47.96`, which is pretty close to the best value of `48.02` we found in our NNI experiment. The next chapters will demonstrate that examining the hyperparameter panel allows you to comprehend many key concepts of an effective deep learning model.

At the bottom of the page is the Trial list panel, which lists all trials. You can observe each trial's parameters, logs, and metrics, as shown in Figure 1-19.

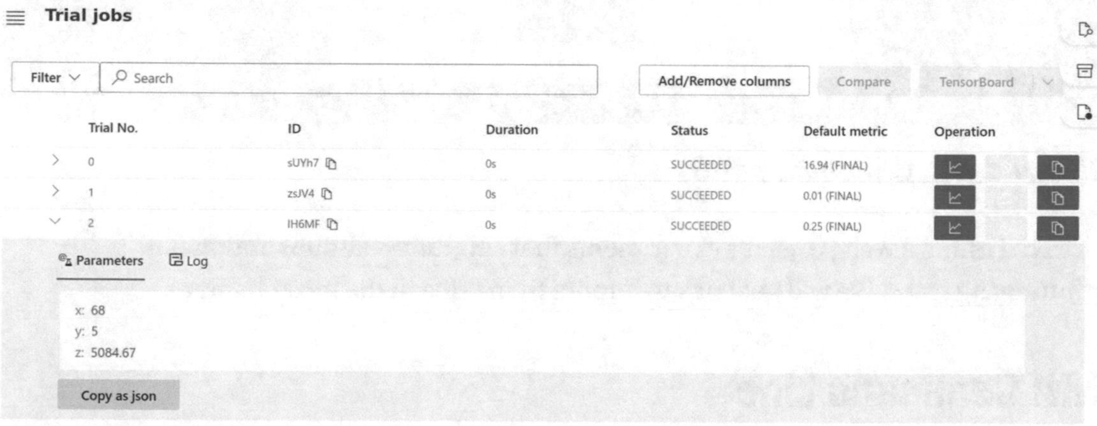

Figure 1-19. *Trial list panel*

An experiment is usually a rather lengthy process that can take days or even weeks. Sometimes, there may be interesting hypotheses to test. For example, it may be necessary to run a trial with specific parameters manually. And if you don't want to wait until the end of the experiment, then you can add a custom trial to the queue by clicking the "Copy" button in the list of challenges. You can enter your trial parameters in the pop-up window and submit a trial. Figure 1-20 demonstrates how you can submit a custom trial.

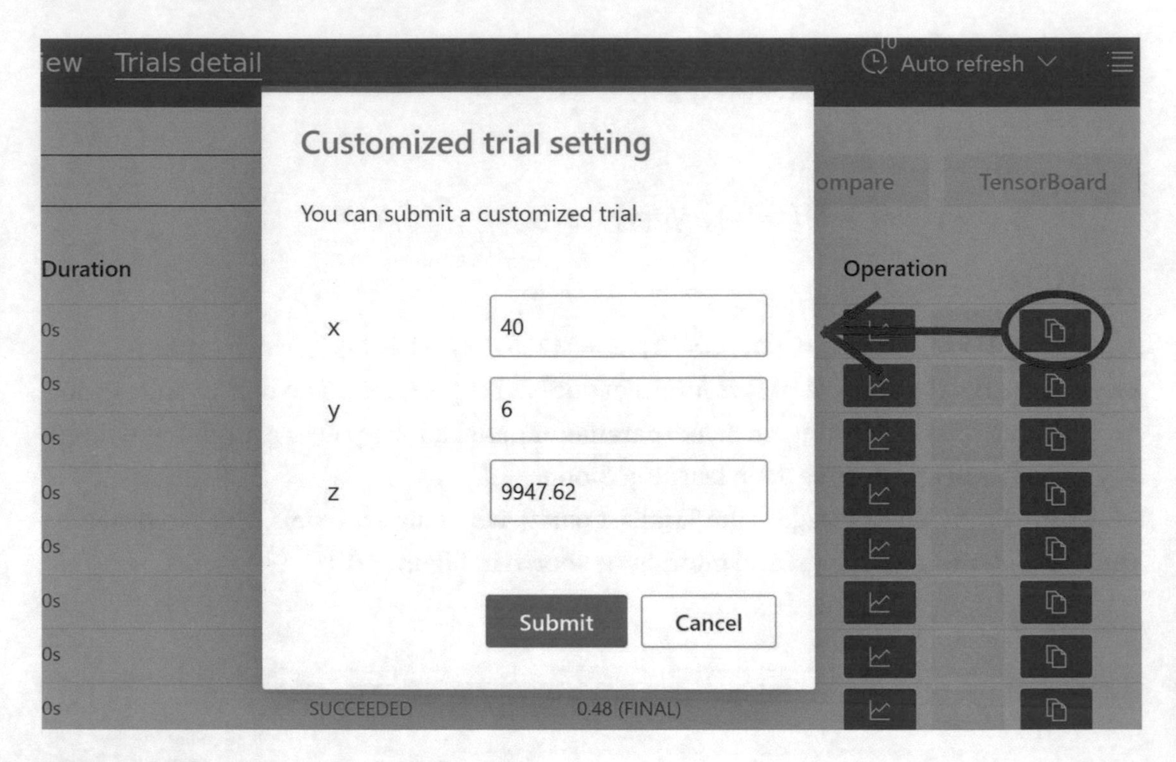

Figure 1-20. *Customized trial*

NNI offers a web panel for experiments that simplifies administration and monitoring tasks. We will get back to it more than once in the next chapters.

NNI Command Line

In addition to the web panel, you can use the command-line interface to manage the NNI and monitor its experiments. NNI has the following working directory: `~/nni-experiments`, where all data about experiments is stored.

`nnictl create --config <path_to_config>`: Starts an experiment and returns `<experiment_id>`. During execution, all information about the running experiment is saved in `~/nni-experiments/<experiment_id>`.

`nnictl stop`: Stops running experiment.

`nnictl experiment list --all`: Returns a list of all created experiments.

`nnictl resume <experiment_id>`: Resumes a stopped experiment. This command is also useful when you want to analyze the results of an already completed experiment.

`nnictl view <experiment_id>`: Outputs information about experiment.

nnictl top: Outputs best trials.

For more information about NNI command-line tool, please refer to

nnictl --help

NNI Experiment Configuration

As we saw earlier, the execution of the experiment is configured through a YAML file (ch1/bbf/config.yml). NNI allows you to configure the experiment execution flexibly. Table 1-2 lists the main configuration parameters.

Table 1-2. *NNI experiment configuration settings*

Field	Type	Description
experimentName	str Optional	Name of the experiment
searchSpaceFile	str Optional	Path to the JSON file containing search space definition
searchSpace	YAML Optional	Field for inline search space definition For example, search space defined in ch1/bbf/search_space.json can be set in searchSpace field: searchSpace: x: _type: quniform _value: [1, 100, 1] y: _type: quniform _value: [1, 10, 1] z: _type: quniform _value: [1, 10000, 0.01]
trialCommand	str Required	Command to execute trial. Use python3 on Linux and macOS, and use python on Windows

(*continued*)

Table 1-2. (*continued*)

Field	Type	Description
trialCodeDirectory	str Optional Default: "."	Path to trial directory
trialConcurrency	int	Number of trials to run concurrently
maxExperimentDuration	str Optional	Limits the experiment duration. Experiment duration is not limited by default Format: number + s|m|h|d Example: maxExperimentDuration: 5h
maxTrialNumber	int Optional	Limits the number of trials. Number of trials is not limited by default
maxTrialDuration	str Optional	Limits the trial duration. Trial duration is not limited by default. Format: number + s|m|h|d Example: maxTrialDuration: 30m
debug	bool Optional Default: False	Enables debug mode
tuner	YAML Optional	Specifies the hyperparameter tuner. Details in *Chapter 3*
assessor	YAML Optional	Specifies the assessor. Details in *Chapter 3*
advisor	YAML Optional	Specifies the advisor. Details in *Chapter 3*
trainingService	YAML Optional	Specifies the training service. Details in *Chapter 7*
sharedStorage	YAML Optional	Specifies the shared storage. Details in *Chapter 7*

For more details, please refer to the official documentation: https://nni.readthedocs.io/en/v2.7/reference/experiment_config.html.

Embedded NNI

Even though the capabilities of the NNI server are quite broad, the NNI can run in embedded mode. It is more convenient to run NNI in Python embedded mode in some cases. This need may arise when it is necessary to dynamically create experiments and have more control over the experiment execution. We will use NNI in embedded mode in some examples in the next chapters.

Listing 1-7 shows an example of the execution of an experiment in embedded mode to optimize the black-box function we examined earlier.

Listing 1-7. Embedded NNI. ch1/bbf/embedded_nni.py

```
# Loading Packages
from pathlib import Path
from nni.experiment import Experiment

# Defining Search Space
search_space = {
    "x": {"_type": "quniform", "_value": [1, 100, 1]},
    "y": {"_type": "quniform", "_value": [1, 10, 1]},
    "z": {"_type": "quniform", "_value": [1, 10000, 0.01]}
}

# Experiment Configuration
experiment = Experiment('local')
experiment.config.experiment_name = 'Black Box Function Optimization'
experiment.config.trial_concurrency = 4
experiment.config.max_trial_number = 1000
experiment.config.search_space = search_space
experiment.config.trial_command = 'python3 trial.py'
experiment.config.trial_code_directory = Path(__file__).parent
experiment.config.tuner.name = 'Evolution'
experiment.config.tuner.class_args['optimize_mode'] = 'maximize'
```

```
# Starting NNI
http_port = 8080
experiment.start(http_port)

# Event Loop
while True:
    if experiment.get_status() == 'DONE':
        search_data = experiment.export_data()
        search_metrics = experiment.get_job_metrics()
        input("Experiment is finished. Press any key to exit...")
        break
```

Listing 1-7 contains an event loop that allows you to track the progress of your experiment automatically. Therefore, you can programmatically design experiments and get the best solutions for a problem.

Troubleshooting

If you experience any problems or errors launching and using NNI, you can follow this mini-guide to determine the issue.

NNI is not starting. In this case, you'll see the error output message after running `nnictl start` command, and this error message can help you fix the problem.

NNI is starting, but you see an ERROR badge in the overview web panel, as shown in Figure 1-21.

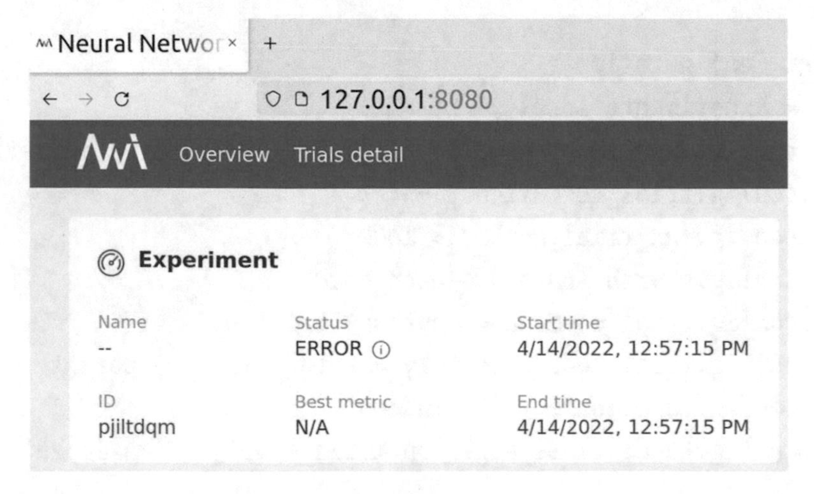

Figure 1-21. *NNI. Error badge*

In this case, please check error log file in ~/nni-experiments/<experiment_id>/log/nnictl_stderr.log.

NNI is starting. Experiment is RUNNING, but Trials have FAILED status, as shown in Figure 1-22.

Trial jobs

Filter | Search | Add/Remove

	Trial No.	ID	Duration	Status
>	0	r1oWM	0s	FAILED
>	1	U1cr6	0s	FAILED
>	2	JDnvG	0s	FAILED
>	3	x2PEu	0s	FAILED

Figure 1-22. *NNI. Failed Trials*

In this case, check Trial logs in Trial jobs panel, as shown in Figure 1-23.

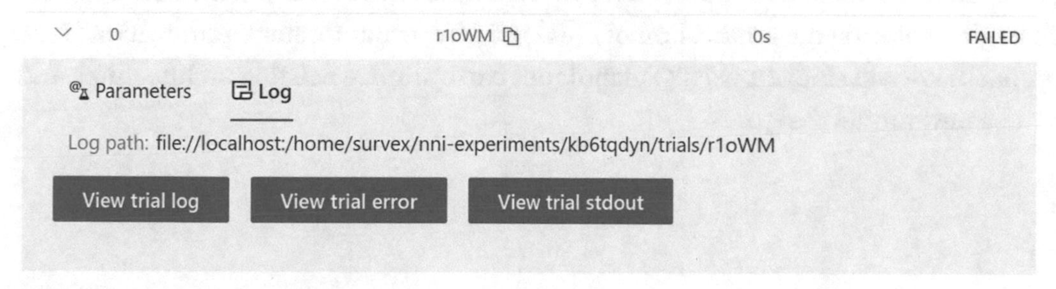

Figure 1-23. *NNI. Trial logs*

This mini-guide may make it easier to find and fix the NNI problem.

TensorFlow and PyTorch

This book will apply AutoDL techniques to models implemented with **TensorFlow** or **PyTorch**. This book assumes that the reader has experience with one of these frameworks. Each chapter will provide examples of applying NNI to a model implemented in TensorFlow or PyTorch. Examples implemented on PyTorch or

TensorFlow will not duplicate each other but will be close to each other. Therefore, if you are only a PyTorch user, you will not lose anything if you do not dive into the examples with TensorFlow models.

This book will use the following framework versions:

- **TensorFlow**: 2.7.0
- **PyTorch**: 1.9.0
- **PyTorch Lightning**: 1.4.2
- **Scikit-learn**: 0.24.1

In any case, I recommend you to go through all examples because their concepts can be easily ported to your favorite deep learning framework.

Summary

In this chapter, we have explored the NNI basic features. NNI is a very powerful toolkit for solving various AutoML tasks. And at the beginning of this chapter, we separately investigated the demand to apply AutoML techniques in practice. In the next chapter, we will begin exploring the application of the classic Hyperparameter Optimization (**HPO**) approach. We will study how HPO techniques can optimize existing architectures and create a new model design.

CHAPTER 2

Hyperparameter Optimization

Almost every deep learning model has a large number of hyperparameters. Choosing the proper hyperparameters is one of the most common problems in AutoML. A small change in one of the model's hyperparameters can significantly change its performance. Hyperparameter Optimization (HPO) is the first and most effective step in deep learning model tuning. Due to its ubiquity, Hyperparameter Optimization is sometimes regarded as synonymous with AutoML.

NNI provides a broad and flexible set of HPO tools. This chapter will examine various neural network designs and how NNI can be applied to optimize their hyperparameters for particular problems.

What Is Hyperparameter?

Let's start the chapter by defining what a model *hyperparameter* is. A deep learning model has three types of parameters:

- **Weights and biases (or model parameters)**: Parameters of linear (or tensor) functions in the neural network's architecture, which are tuned during its training.
- **Hyperparameters**: Initial global variables that are set manually and affect the behavior of the functions, the training algorithm, and the neural network's architecture.

I. Gridin, *Automated Deep Learning Using Neural Network Intelligence*,
https://doi.org/10.1007/978-1-4842-8149-9_2

- **Task parameters**: The parameters that the task sets for you. These parameters lie in the problem requirements, which need to be satisfied and cannot be changed. For example, suppose we solve the binary classification problem determining "cat or dog?" by analyzing their pictures. In that case, we have a task parameter: **2**, which indicates the number of output classes. Or, for example, we have the air temperature prediction problem for the next three days. Then parameter 3 is a task parameter, and it lies in the task requirements and cannot be changed in any way.

Let's look at an example of a Fully Connected Neural Network (or Dense Network) with three linear (or dense) layers with activation functions and five-valued input vector and scalar output, which can be represented as follows in the TensorFlow framework:

We import necessary packages:

Listing 2-1. Fully Connected Neural Network. ch2/hpo_definition/fcnn_model.py

```
import tensorflow as tf
from tensorflow.keras.layers import Dense
```

Next, we set task parameters which are task requirements. Our Fully Connected Neural Network has to receive five-valued input vector and output a scalar value:

```
# Task Parameters
inp_dim = 5
out_dim = 1
```

Since we have three linear (or dense) layers, we can specify the `output_dimension` value for two of them. The third layer has an `output_dimension` value of 1 because this is a task requirement. These values are hyperparameters:

```
# Hyperparameters
l1_dim = 8
l2_dim = 4
```

We initialize the FCNN model:

```
# Model
model = tf.keras.Sequential(
    [
        Dense(l1_dim, name = 'l1',
              activation = 'sigmoid', input_dim = inp_dim),
        Dense(l2_dim, name = 'l2',
              activation = 'relu'),
        Dense(out_dim, name = 'l3'),
    ]
)
model.build()
```

And here we have FCNN model parameters:

```
# Weights and Biases
print(model.summary())
```

which are presented in Table 2-1.

Table 2-1. *FCNN model weights and biases*

Layer	Output Shape	Param #	Explained
l1	(None, 8)	48	5×8 weight matrix + 8 bias vector = 48
l2	(None, 4)	36	8×4 weight matrix + 4 bias vector = 36
l3	(None, 1)	5	4×1 weight matrix + 1 bias vector = 5

```
Total params: 89
```

The model shown in Listing 2-1 has 2 hyperparameters and 89 model parameters. Hyperparameters usually directly affect the number of model parameters. Indeed, in Listing 2-1, the `l1_dim` and `l2_dim` hyperparameters set the dimensions of weight matrices and bias vectors. Figure 2-1 illustrates the hyperparameter impact on the FCNN model architecture and its parameters.

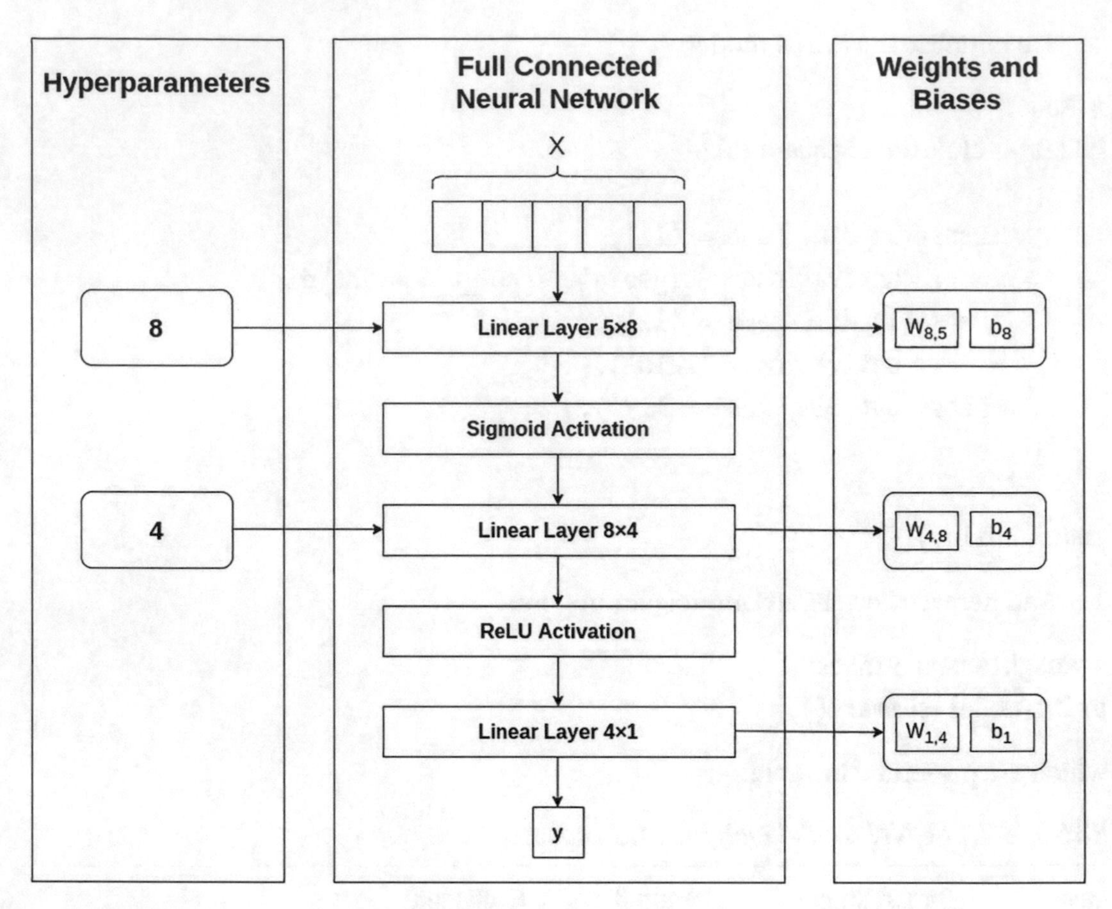

Figure 2-1. *Hyperparameter impact*

We can distinguish four types of hyperparameters:

- ***Layer*** hyperparameter
- ***Training*** hyperparameter
- ***Feature*** hyperparameter
- ***Design*** hyperparameter

Let's examine each of these hyperparameter types.

Layer Hyperparameter

Almost all layers of a deep learning model imply the presence of initial parameters. For example:

- **Dropout layer**: Assumes p $(0 < p < 1)$ parameter, which defines dropout probability
- **MaxPool 2D layer**: Assumes `pool_size` parameter, which defines the pooling dimension
- **Convolutional 2D layer**: Assumes `kernel_size` parameter

We can refer to these hyperparameters as *layer hyperparameters*.

Training Hyperparameter

The training process is an integral part of the model architecture. Each model generates a multidimensional loss function surface. The model training process tries to find the best local minima on the loss function surface. The training process parameters can drastically affect trained model performance.

The most common example is *learning rate* tuning. Most training algorithms use gradient descent as the main idea behind model training. The gradient descent concept means that a transition vector is calculated for each point on the loss function surface. But the length of this vector is determined by the *learning rate* parameter. Too high *learning rate* parameter can lead to a gradient descent explosion and a complete inability to find an acceptable local minima on the loss function surface. At the same time, too low *learning rate* stops the training process at a too high point on the surface and does not allow model parameters to reach a lower point. Figure 2-2 demonstrates the learning rate problem.

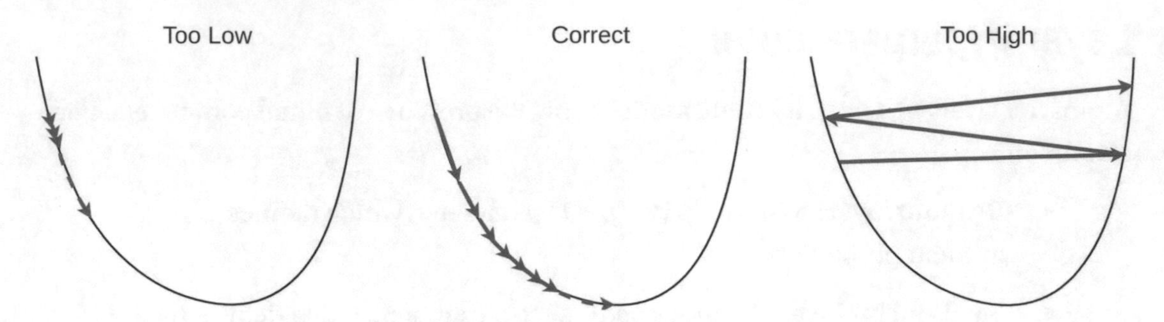

Figure 2-2. *Learning rate problem*

The most common training hyperparameters are

- *Training epochs number*
- *Learning rate*
- *Batch size*

Layer Hyperparameter and Training Hyperparameter Optimization is the most common way to tune a model due to the ease of this approach implementation.

Feature Hyperparameter

Feature hyperparameter affects dataset preprocessing methods. The data structure in the input dataset can significantly improve the model's performance, especially in natural language processing (NLP) problems. But transformations of the input dataset do not always improve the model's performance, so you often have to "play" with various feature preprocessing techniques to reach the best results.

Let's examine a dataset that contains movie reviews data. This dataset includes the following features:

- **Feature *A***: Movie budget (*example: 100 000 000$*)
- **Feature *B***: Review date (*example: 2021-05-03*)
- **Feature *C***: Review text (*example: I love good movies, but unfortunately, this is not one of them.*)

And we are solving the classical binary classification problem, that is, we have to determine whether the review is negative or positive. Then we can apply the following preprocessing, which is shown in Figure 2-3.

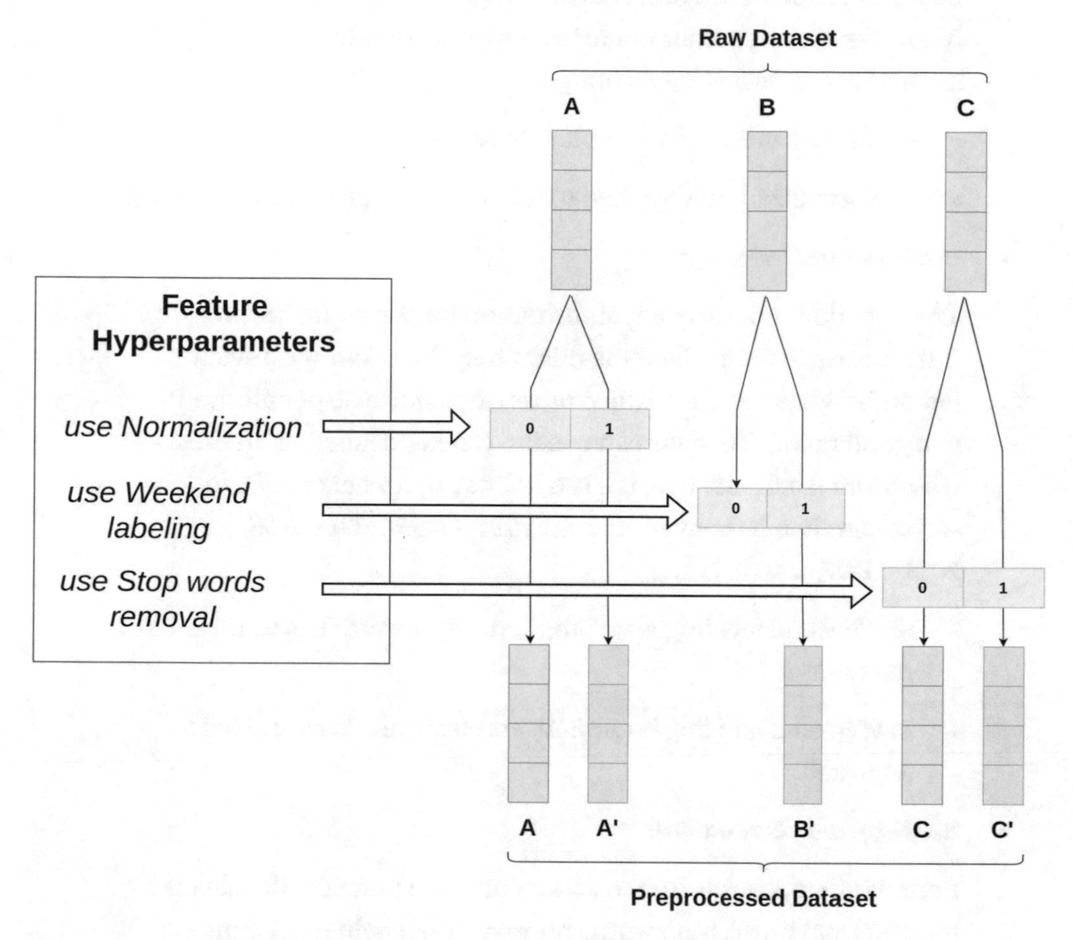

Figure 2-3. *Dataset preprocessing driven by feature hyperparameters*

The dataset preprocessing shown in Figure 2-3 has the following feature hyperparameters: *use Normalization, use Weekend labeling,* and *use Stop words removal.* Let's describe their meanings:

- ***use Normalization:***

  Normalization is a common technique for converting numeric data by converting all values to the range from 0 to 1. *use Normalization* hyperparameter manages the application of normalization to feature A (budget):

  - **0**: Normalization is not applied to feature A.
  - **1**: Normalization is applied to feature A and produces A` feature.

- ***use Weekend labeling:***

  The date does not carry any information for the neural network. But an extra datetime labeling might help. For example, reviews left on holidays can be positive more often because people are in a good mood. Then we can use the weekend labeling method, which will label each date if it is a holiday or a weekend. Date series can then be converted from *2021-11-05, 2021-11-06, 2021-11-07, ...* to 0, 1, 1, ...

  - **0**: Weekend labeling is not applied, and feature B is removed from the dataset.
  - **1**: Weekend labeling is applied, and feature B is converted to feature B′.

- ***use Stop words removal:***

  Removing stop words from text is a common practice that cleans the text from noise. Stop words removal often helps speed up training and improve the quality of an NLP model.

  - **0**: Stop words removal is not applied to feature C.
  - **1**: Stop words removal is applied to feature C and produces feature C`.

For example, this combination of feature hyperparameters {*use Normalization*: 0, *use Weekend labeling*: 1, *use Stop words removal*: 1} will transform original dataset [A, B, C] to [A, B′, C′].

Design Hyperparameter

Design hyperparameter has a direct impact on the choice of neural network architecture. Their values control the choice of neural network layers and connections between them.

Figure 2-4 shows design hyperparameters.

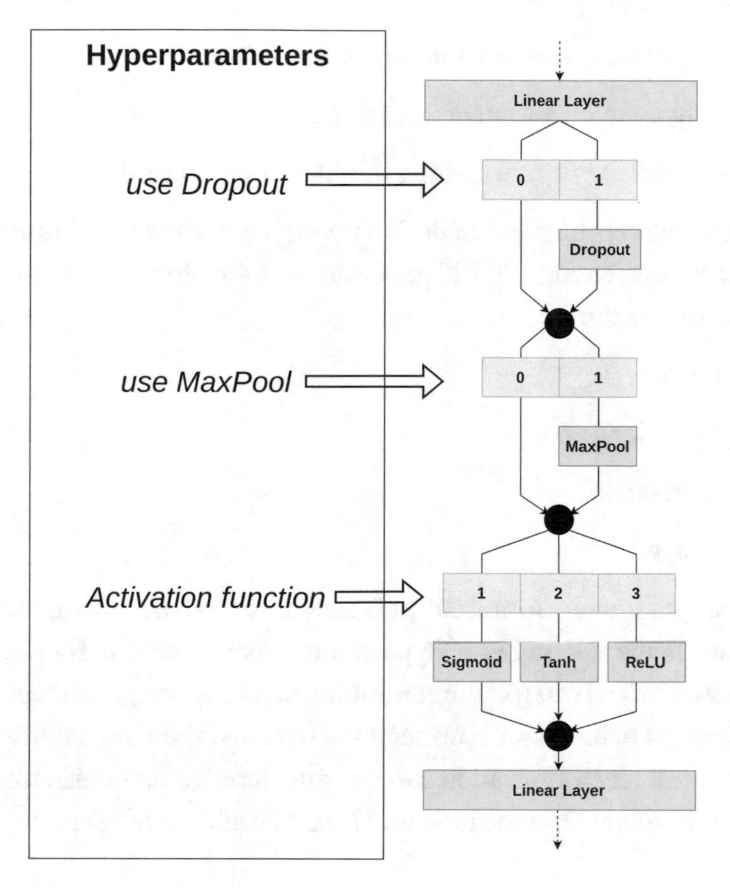

Figure 2-4. *Design hyperparameters*

The design hyperparameter search shown in Figure 2-4 has the following design hyperparameters: *use Dropout, use MaxPool,* and *Activation function*. And they affect the model design the following way:

- ***use Dropout:***
    - **0**: *Dropout layer* is skipped.
    - **1**: *Dropout layer* is connected.

- ***use MaxPool:***
  - **0**: *MaxPool layer* is skipped.
  - **1**: *Max Poll layer* is connected.
- ***Activation function:***
  - **1**: *Sigmoid activation* function is connected.
  - **2**: *Hyperbolic tangent activation* function is connected.
  - **3**: *Rectified linear activation* function is connected.

For example, this combination of design hyperparameters {*use Dropout*: 1, *use MaxPool*: 0, *Activation function*: 3} will generate the following sequence of layers in the neural network architecture:

- Linear layer
- Dropout layer
- ReLU activation
- Linear layer

Design Hyperparameter Optimization is used less often because it is more challenging to implement than Layer Hyperparameter or Training Hyperparameter Optimization. But design hyperparameter tuning can produce great results. It chooses the best combination of layers and connections between them for a particular problem. Design Hyperparameter Optimization can be considered as an intermediate approach between Hyperparameter Optimization and Neural Architecture Search.

Search Space

Say we have determined the model's hyperparameters, which will need to be optimized. Next, we must define a search space for each of the hyperparameters. Determining the search space requires some experience and intuition. You must understand that the larger the search space, the longer the experiment. And it is more difficult to find a suitable solution. Therefore, it is pointless to specify a huge number of values in the search space. For example, if `l1` is a hyperparameter that specifies the dimension of a linear layer (`tensorflow.keras.layers.Dense(l1)` or `torch.nn.Linear(out_features = l1)`), then you don't need to set the search space for the hyperparameter to

`[1, 2, 3,..., 999, 1000]`. For the hyperparameter `l1`, values expressing the power of two (2^n) are more suitable: `[4, 8, 16, ..., 256]`, because the representation size of linear layers is relevant only proportionally, not additively (if `l1 = 256` performed poorly, thus, it is highly likely that `l1 = 256 + 8` will show the same result). A reasonable choice of hyperparameters can significantly reduce the time of the experiment without losing its quality. Before specifying a search space, you can perform manual exploration to determine which hyperparameter values have the most impact on model performance.

The search space for the HPO problem is defined by defining a set of possible values for each of the hyperparameters. NNI allows the following sampling strategies to define hyperparameter search space: `choice`, `randint`, `uniform`, `quniform`, `loguniform`, `qloguniform`, `normal`, `qnormal`, `lognormal`, and `qlognormal`.

choice

```
{"_type": "choice", "_value": options}
```

Choice sampling strategy allows you to manually specify a list of values that a hyperparameter can take. It can be a list of numbers and strings. For example:

```
"hp": {"_type": "choice", "_value": [128, 512, 1024]}
```

Choice sampling also supports nested search spaces. Nested choice is especially useful when dealing with design hyperparameters. Here is the example of nested choice sampling:

```
"layer1":{
  "_type": "choice",
  "_value": [{"_name": "Empty"},
    {
      "_name": "Conv", "kernel_size":
      {"_type": "choice", "_value": [1, 2, 3, 5]}
    },
    {
      "_name": "Max_pool", "pooling_size":
      {"_type": "choice", "_value": [2, 3, 5]}
    },
```

```
    {
      "_name": "Avg_pool", "pooling_size":
      {"_type": "choice", "_value": [2, 3, 5]}
    }
  ]
}
```

randomint

```
{"_type": "randint", "_value": [lower, upper]}
```

Chooses random integer from lower (inclusive) to upper (exclusive).

uniform

```
{"_type": "uniform", "_value": [low, high]}
```

Chooses random value according to uniform distribution on `[low, high]`.

quniform

```
{"_type": "quniform", "_value": [low, high, q]}
```

Acts like uniform sampling but with `q` discretization that can be expressed as `clip(round(uniform(low, high) / q) * q, low, high)`. For example, for _value specified as `[1, 11, 2.5]`, possible values are `[1, 2.5, 5, 7.5, 10, 11]`.

loguniform

```
{"_type": "loguniform", "_value": [low, high]}
```

Chooses random value according to loguniform distribution on `[low, high]` that can be expressed as `np.exp(uniform(np.log(low), np.log(high)))`.

qloguniform

```
{"_type": "qloguniform", "_value": [low, high, q]}
```

Acts like loguniform sampling but with q discretization that can be expressed as `clip(round(loguniform(low, high) / q) * q, low, high)`.

normal

```
{"_type": "normal", "_value": [mu, sigma]}
```

Chooses random value according to normal distribution with μ = `mu` and σ = `sigma`.

qnormal

```
{"_type": "qnormal", "_value": [mu, sigma, q]}
```

Acts like normal sampling but with q discretization that can be expressed as `round(normal(mu, sigma) / q) * q`.

lognormal

```
{"_type": "lognormal", "_value": [mu, sigma]}
```

Chooses random value according to lognormal distribution with μ = `mu` and σ = `sigma` that can be expressed as `np.exp(normal(mu, sigma))`.

qlognormal

```
{"_type": "qlognormal", "_value": [mu, sigma, q]}
```

Acts like lognormal sampling but with q discretization that can be expressed as `round(exp(normal(mu, sigma)) / q) * q`.

The implementation of sampling strategies is in `nni.parameter expressions`. You can explore search space sampling strategies manually, as shown in Listing 2-2.

Listing 2-2. quniform sampling strategy. ch2/search_space/quniform.py

```
import nni
from numpy.random.mtrand import RandomState
import matplotlib.pyplot as plt
```

We generate 20 samples using quniform strategy:

```
space = [
    nni.quniform(0, 100, 5, RandomState(seed))
    for seed in range(20)
]
```

Visualize generated samples:

```
plt.figure(figsize = (5, 1))
plt.title('quniform')
plt.plot(space, len(space) * [0], "x")
plt.yticks([])
plt.show()
```

And after, you can observe generated samples using `quniform` method in Figure 2-5.

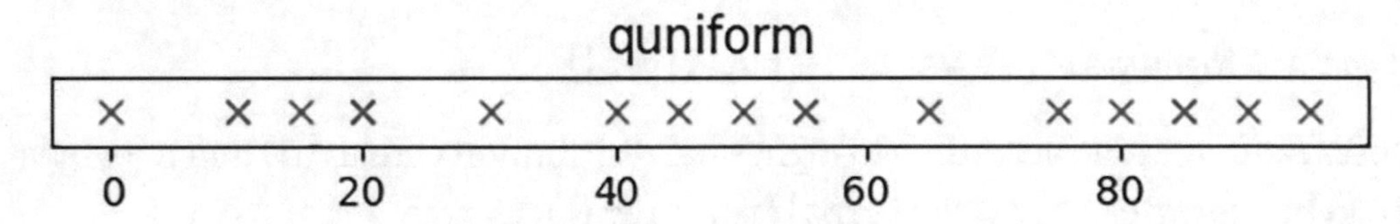

Figure 2-5. *quniform samples*

Let's examine an example of a JSON search space definition.

Listing 2-3. Search space. ch2/search_space/search_space.json

```
{
  "dropout_rate":
  { "_type": "uniform", "_value": [0.1, 0.5]},

  "conv_size":
  {"_type": "choice", "_value": [2, 3, 5, 7]},
```

```
    "layer1_hidden_size":
    {"_type": "choice", "_value": [128, 512, 1024]},

    "layer2_hidden_size":
    {"_type": "choice", "_value": [16, 32, 64]},

    "activation_function":
    {"_type": "choice", "_value": ["tanh", "sigmoid", "relu"]},

    "training_batch_size":
    {"_type": "choice", "_value": [100, 250, 500]},

    "training_learning_rate":
    {"_type": "uniform", "_value": [0.0001, 0.1]}
}
```

Listing 2-3 demonstrates a typical search space for deep learning model:

- **dropout_rate**: Layer hyperparameter that defines the *p* parameter in dropout layer. `dropout_rate` can take any value from 0.1 to 0.5.
- **conv_size**: Layer hyperparameter that defines the kernel size of convolutional layer. `conv_size` can take any value from the list: 2, 3, 5, 7.
- **layer1_hidden_size**: Layer hyperparameter that defines the output dimension of the first linear layer. `layer1_hidden_size` can take any value from the list: 128, 512, 1024.
- **layer2_hidden_size**: Layer hyperparameter that defines the output dimension of the second linear layer. `layer2_hidden_size` can take any value from the list: 16, 32, 64.
- **activation_function**: Design hyperparameter that defines output activation function. `activation_function` can take any value from the list: tanh, sigmoid, relu.
- **training_batch_size**: Training hyperparameter that defines batch size that will be used during training. `training_batch_size` can take any value from the list: 100, 250, 500.
- **training_learning_rate**: Training hyperparameter that defines learning rate. `training_learning_rate` can take any value from 0.0001 to 0.1.

Tuners

After defining the search space, we need to define a tuner that will explore the search space and select trial hyperparameter combinations based on the existing results.

The tuner is set as follows in the configuration file:

```
tuner:
  name: <Tuner_Name>
  classArgs:
      optimize_mode: minimize
      arg1: val1
      arg2: val2
```

Each tuner has its own set of parameters. The only common parameter for all tuners is `optimize_mode`, which marks the direction of optimization of the metric that characterizes the model's performance: `minimize`, `maximize`.

Table 2-2 provides the list of tuners available in NNI v2.7.

Table 2-2. *Search Tuners*

Configuration Id	Name
SMAC	Sequential Model-Based Optimization
TPE	Tree-structured Parzen Estimator
Random	Random Search
Anneal	Annealing Search Algorithm
Evolution	Genetic Algorithm Search
BatchTuner	Batch Tuner
GridSearch	Grid Search
NetworkMorphism	Network Morphism
MetisTuner	Metis Tuner
GPTuner	Gaussian Process (GP) Tuner
PBTTuner	Population-Based Training Tuner

We will study tuning algorithms in detail in Chapter 3. In this chapter, we will consider only **Random Search Tuner** and **Grid Search Tuner**.

Random Search Tuner

The Random Search Tuner is the most straightforward approach to choosing a combination of hyperparameters. As the name implies, the combination of hyperparameters is chosen absolutely randomly. Despite the simplicity of this approach, it can sometimes give very good results.

Random Search Tuner is set as

```
tuner:
  name: Random
```

Figure 2-6 illustrates the Random Search Tuner in action.

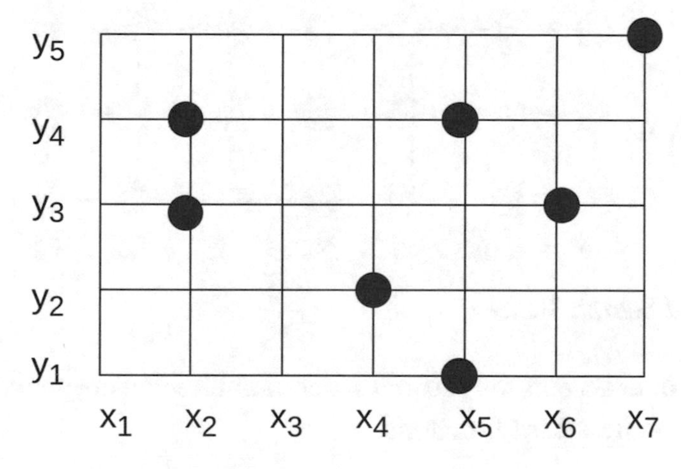

Figure 2-6. *Random Search Tuner*

In some cases, Random Search Tuner is well suited for exploring the search space when you need to extract information about hyperparameters' impact on model performance. After a random search space exploration, you can redefine hyperparameter search space and select another tuner.

Grid Search Tuner

Grid Search Tuner performs an exhaustive search, that is, Grid Search Tuner will search all the possible combinations from the search space. Grid Search Tuner is well suited for small search spaces.

Grid Search Tuner is set as

```
tuner:
  name: GridSearch
```

Figure 2-7 illustrates the Grid Search Tuner in action.

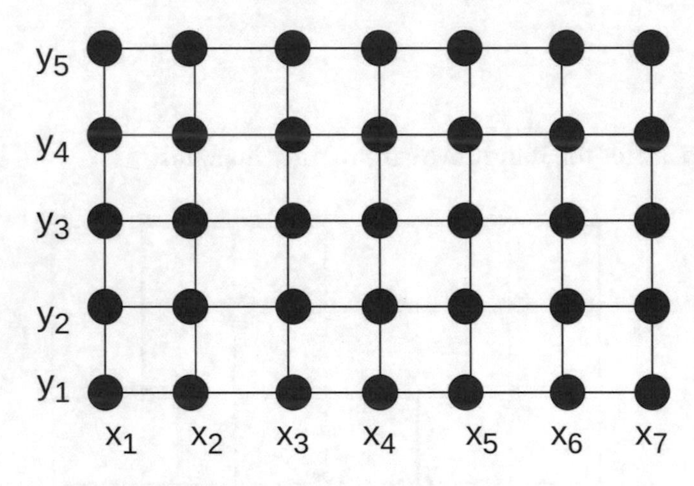

Figure 2-7. *Grid Search Tuner*

Grid Search Tuner accepts only search space variables that are generated with `choice`, `quniform`, and `randint` functions.

Organizing Experiment

And so, we are all set to begin our first explorations. Let's look at the file organization pattern we'll be using in this book. I also recommend that you follow the same approach.

These simple rules will help you avoid unnecessary errors when running an experiment:

- Use a separate directory for each experiment.
- Specify the current directory as trial code directory in the experiment configuration.

- Keep the model class and training/testing methods in a separate file.
- Add root code folder to the system path in trial file.
- Do not mix model and trial files.

Let's look at a dummy experiment that follows these rules ch2/experiment_pattern. Listing 2-4 provides the experiment configuration file.

The configuration file marks the current directory as the working directory: `trialCodeDirectory: .`

Listing 2-4. Experiment configuration. ch2/experiment_pattern/config.yml

```
trialConcurrency: 1
searchSpaceFile: search_space.json
trialCodeDirectory: .
trialCommand: python3 trial.py
tuner:
  name: GridSearch
trainingService:
  platform: local
```

Model class and training/testing methods are in a separate file shown in Listing 2-5.

Listing 2-5. DummyModel class. ch2/experiment_pattern/model.py

```
from random import random

class DummyModel:

    def __init__(self, x, y) -> None:
        super().__init__()
        self.x = x
        self.y = y

    def train(self):
        # Training here
        ...

    def test(self):
        # Test results
        return round(self.x + self.y + random() / 10, 2)
```

Table 2-3. *NNI API*

Method	Description
`nni.get_next_parameter()`	Required method that receives trial parameters from NNI Tuner as `Dict` object
`nni.report_intermediate_result(m)`	Optional method that sends intermediate results to NNI Tuner
`nni.report_final_result(m)`	Required method that sends final metrics that represents model's performance

The trial script receives parameters from the NNI tuner, initializes the model, trains it, and tests its performance. The trial script interacts with NNI using the following API methods shown in Table 2-3.

Let's look at the trial script pattern in Listing 2-6.

We import necessary modules:

Listing 2-6. Trial script pattern. ch2/experiment_pattern/trial.py

```
import os
import sys
import nni
```

And here, we add the root directory of the code to the system path. This is done because NNI has no idea about the structure of our code and the location of modules.

```
# For NNI use relative import for user-defined modules
SCRIPT_DIR = os.path.dirname(os.path.abspath(__file__)) + '/../..'
sys.path.append(SCRIPT_DIR)
```

Now, we can import the classes we need from our code structure:

```
from ch2.experiment_pattern.model import DummyModel
```

Trial initiates the model, trains it, measures its performance, and returns the result to NNI Tuner:

```
def trial(hparams):
    """
    Trial Script:
      - Initiate Model
      - Train
      - Test
      - Report
    """
    model = DummyModel(**hparams)
    model.train()
    accuracy = model.test()

    # send final accuracy to NNI
    nni.report_final_result(accuracy)
```

And here is the entry point to the trial script:

```
if __name__ == '__main__':

    # Manual HyperParameters
    hparams = {
        'x': 1,
        'y': 1,
    }

    # NNI HyperParameters
    # Run safely without NNI Experiment Context
    nni_hparams = nni.get_next_parameter()
    hparams.update(nni_hparams)

    trial(hparams)
```

You can run trial script ch2/experiment_pattern/trial.py in stand-alone mode. It means you can run this script without getting any errors. The `nni.get_next_parameter()` method will return an empty `dict` that you can merge with your trial parameters. This is convenient if you want to test trial execution.

Experiment file structure pattern can be depicted as it is shown in Figure 2-8.

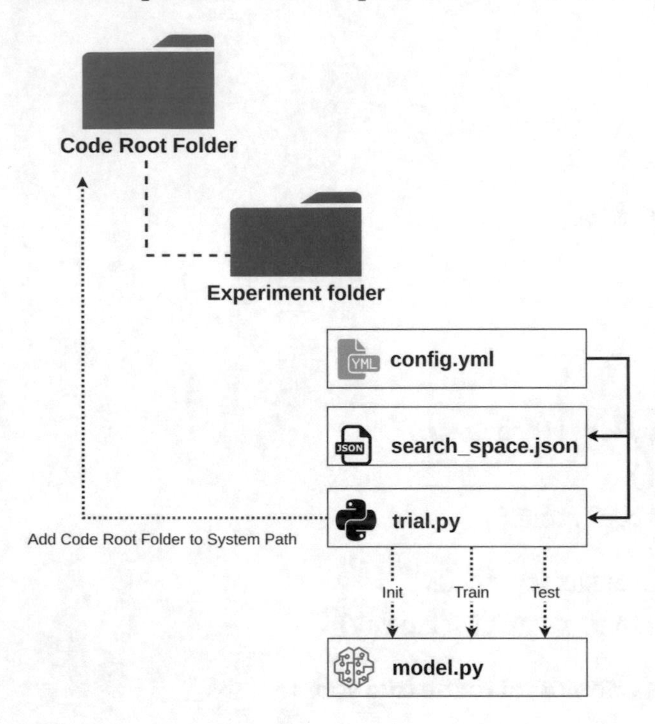

Figure 2-8. *Experiment file structure*

Fine! We've defined different hyperparameter types, examined how to define search spaces, studied simple tuners, and represented a pattern for creating experiments. We are now ready to move on to real research. The following sections will examine how HPO methods optimize the model for the specific problem and help develop a new model design.

Optimizing LeNet for MNIST Problem

Libraries: TensorFlow (Keras API), PyTorch

MNIST Classifier problem is a common problem to test various machine learning approaches. Convolutional neural networks have made a breakthrough in image recognition, so we also use the MNIST dataset for diving into the HPO area. The MNIST database contains images of handwritten digits. It is split into two sets: a training set of 60,000 samples and a test set of 10,000 examples. Figure 2-9 displays several samples from the MNIST dataset.

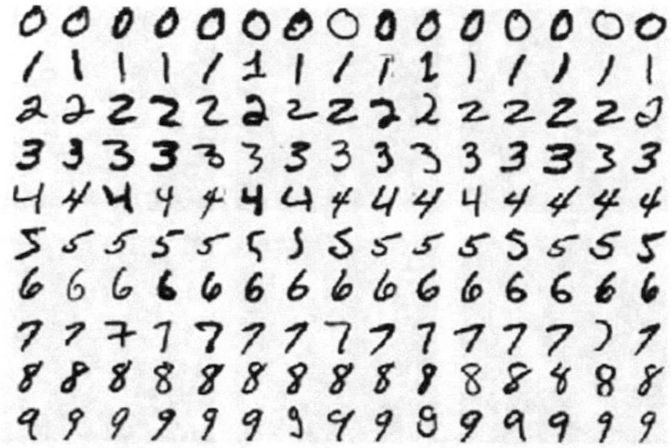

Figure 2-9. *MNIST database*

MNIST dataset is a set of 28×28 grayscale images. Therefore, each dataset object is a (28, 28, 1) tensor. Let's examine several samples of this dataset.

Listing 2-7. MNIST dataset samples. ch2/lenet_hpo/mnist_dataset.py

```
import tensorflow_datasets as tfds

ds, info = tfds.load('mnist', split = 'train', with_info = True)
fig = tfds.show_examples(ds, info)
fig.show()
```

Listing 2-7 displays the image in Figure 2-10.

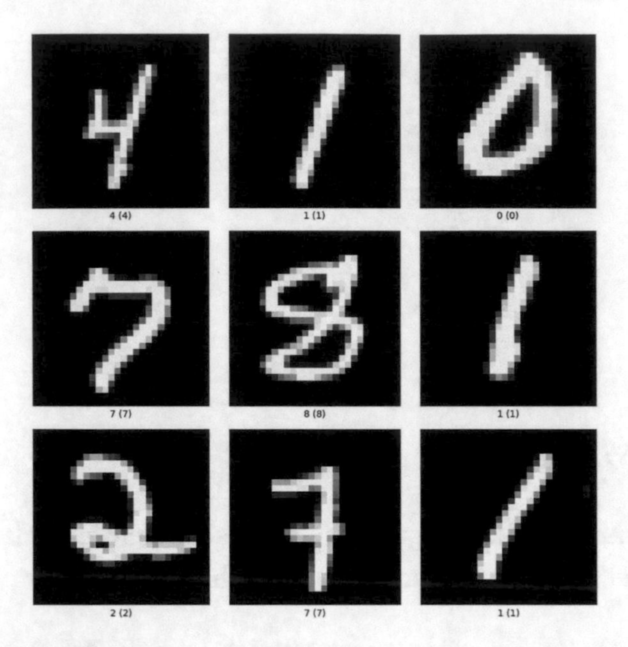

Figure 2-10. *MNIST samples*

This may seem like a very simple task, but it is not. Recall how often you could not understand the number written down by another person's hand. Handwritten digit recognition was one of the first fundamental problems of pattern recognition. LeNet-5 is one of the earliest neural networks used for recognizing handwritten and machine-printed characters. The main reason behind the popularity of this model was its straightforward architecture. It is a multilayer convolution neural network for image classification. Abstract LeNet model design is depicted in Figure 2-11.

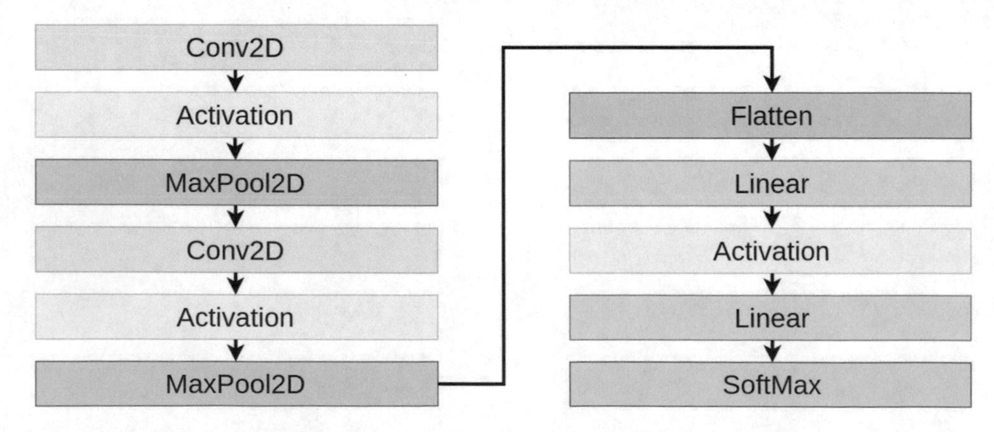

Figure 2-11. *LeNet architecture*

Our goal is to optimize the LeNet model to the handwritten digit recognition problem. The easiest thing is to start Layer Hyperparameter Optimization. Let's just count the number of layer hyperparameters in the LeNet model:

- **Conv2D layer**: Five basic hyperparameters (`out_channels`, `kernel_size`, `stride`, `padding`, `dilation`)
- **MaxPool2D layer**: Two basic hyperparameters (`kernel_size`, `stride`)
- **Linear layer:** Two basic hyperparameters (`out_features`, `use_bias`)

LeNet contains two Conv2D layers, two MaxPool2D layers, and two linear layers. Thus, we have 2×5 + 2×2 + 2×2 = 14 layer hyperparameters in LeNet model. Let's assume that for each hyperparameter, we will have a set of possible values that will consist of only two elements, although, of course, many hyperparameters require more values for the flexibility of the experiment. But even this binary search space contains $2^{14} = 16\,384$ elements. And these are just the most primitive layer hyperparameters for one of the simplest deep learning models. As the complexity of the model increases, the number of possible hyperparameters in it grows exponentially. Even the most advanced tuner can take a very long time to explore this search space. Therefore, we need some experience and intuition, which will allow us to select the critical hyperparameter range for each model without increasing the search space too much.

Let's consider only the following layer hyperparameters:

- **Conv2D layer 1**: `filter_size_1`, `kernel_size_1`
- **Conv2D layer 2**: `filter_size_2`, `kernel_size_2`
- **Linear layer**: `out_features`

In order to reduce the size of the search space, we set

- `filter_size_2 = 2 * filter_size_1`
- `kernel_size_1 = kernel_size_2`

Then, we will focus on only three hyperparameters for LeNet model:

- `filter_size`
- `kernel_size`
- `out_features`

Usually, the best values for the `filter_size` of the convolutional layer are presented as 2^n. Therefore, we will choose the following ones for the search space: 8, 16, 32. The `kernel_size` values are usually chosen in a set of 2n + 1. And the larger the image is, the larger `kernel_size` value should be chosen. The samples of the MNIST dataset are 28×28 images. These are pretty small images, so we shouldn't choose large `kernel_size` values: 2, 3, 5. The best values for `l1_size` are powers of two, as for `filter_size`. The linear layer is applied to the tensor after the flatten layer, which means that we have to consider the dimension of the tensor that the previous layers will produce. In the case of the MNIST problem, we will focus on the following `l1_size` values: 32, 64, 128. For MaxPool2D layer we will set the lowest possible value of `pool_size`, which is 2, and set `sigmoid` as the activation function. Yes! That's the function that has been used for quite a long time as an activation function for most pattern recognition problems. We will return to the problem of choosing an activation function in the next section.

We will use classic batch neural network training with `Adam` optimizer. So let's now look at training hyperparameters. I suggest starting the study by selecting the simplest hyperparameters: `batch_size` and `learning rate`. The best parameters for `batch_size` are expressed as 2^n and `learning_rate` as 10^{-n}. For this case, we will choose the following: 256, 512, 1024 for `batch_size` and 0.01, 0.001, 0.0001 for `learning_rate`.

Now let's convert the hyperparameter constraints we made earlier into NNI search space. Listing 2-8 defines the search space for LeNet hyperparameter optimization search space.

Listing 2-8. LeNet HPO search space. ch2/lenet_hpo/search_space.json

```
{
  "filter_size": {
    "_type": "choice", "_value": [8, 16, 32]},
  "kernel_size": {
    "_type": "choice", "_value": [2, 3, 5]},
  "l1_size": {
    "_type": "choice", "_value": [32, 64, 128]},
  "batch_size": {
    "_type": "choice", "_value": [256, 512, 1024]},
  "learning_rate": {
    "_type": "choice", "_value": [0.01, 0.001, 0.0001]}
}
```

Fine! In the next step, we will create TensorFlow and PyTorch implementations of the LeNet model considering the HPO problem.

TensorFlow LeNet Implementation

In this section, we will study the LeNet model's implementation using TensorFlow (Keras API). Listing 2-9 demonstrates the implementation of the TensorFlow LeNet model for the hyperparameter optimization.

We import necessary modules:

Listing 2-9. LeNet. TensorFlow implementation. ch2/lenet_hpo/tf_lenet_model.py

```
from tensorflow.keras import Model
from tensorflow.keras.layers import Conv2D, Dense, Flatten, MaxPool2D
from tensorflow.keras.optimizers import Adam
from ch2.utils.datasets import mnist_dataset
from ch2.utils.tf_utils import TfNniIntermediateResult
```

Next, we define the LeNet model with three layer hyperparameters:

```
class TfLeNetModel(Model):

    def __init__(self, filter_size, kernel_size, l1_size):
        super().__init__()
```

First convolutional stack:

```
        self.conv1 = Conv2D(
            filters = filter_size,
            kernel_size = kernel_size,
            activation = 'sigmoid'
        )
        self.pool1 = MaxPool2D(pool_size = 2)
```

Second convolutional stack:

```
        self.conv2 = Conv2D(
            filters = filter_size * 2,
            kernel_size = kernel_size,
```

```
            activation = 'sigmoid'
        )
        self.pool2 = MaxPool2D(pool_size = 2)
```

Dense stack:

```
        self.flatten = Flatten()
        self.fc1 = Dense(
            units = l1_size,
            activation = 'sigmoid'
        )
        self.fc2 = Dense(
            units = 10,
            activation = 'softmax'
        )
```

LeNet is a straightforward model which passes calculation results from one layer to another:

```
    def call(self, x, **kwargs):
        x = self.conv1(x)
        x = self.pool1(x)
        x = self.conv2(x)
        x = self.pool2(x)
        x = self.flatten(x)
        x = self.fc1(x)
        return self.fc2(x)
```

The training method uses two training hyperparameters: `batch_size` and `learning rate`. We use `Adam` optimizer with categorical cross-entropy loss function:

```
    def train(self, learning_rate, batch_size):
        self.compile(
            optimizer = Adam(learning_rate = learning_rate),
            loss = 'sparse_categorical_crossentropy',
            metrics = ['accuracy']
        )
        (x_train, y_train), _ = mnist_dataset()
```

Next, we initialize a callback that sends intermediate results to NNI:

```
intermediate_cb = TfNniIntermediateResult('accuracy')
```

Performing classic batch training with ten epochs:

```
self.fit(
    x_train,
    y_train,
    batch_size = batch_size,
    epochs = 10,
    verbose = 0,
    callbacks = [intermediate_cb]
)
```

And the last method we need to define is model testing. We load the test MNIST dataset and perform the classification by measuring its accuracy:

```
def test(self):
    """Testing Trained Model Performance"""
    (_, _), (x_test, y_test) = mnist_dataset()
    loss, accuracy = self.evaluate(x_test, y_test, verbose = 0)
    return accuracy
```

Well, since the implementation of TensorFlow LeNet model is ready, we can implement the NNI trial script using Listing 2-10.

We import necessary modules and pass code root directory to system path:

Listing 2-10. NNI trial script with TensorFlow LeNet implementation. ch2/lenet_hpo/tf_trial.py

```
import os
import sys
import nni

# For NNI use relative import for user-defined modules
SCRIPT_DIR = os.path.dirname(os.path.abspath(__file__)) + '/../..'
sys.path.append(SCRIPT_DIR)

from ch2.lenet_hpo.tf_lenet_model import TfLeNetModel
```

The trial method initializes the model, trains it, tests it, and returns the NNI metric:

```
def trial(hparams):
    model = TfLeNetModel(
        filter_size = hparams['filter_size'],
        kernel_size = hparams['kernel_size'],
        l1_size = hparams['l1_size']
    )
    model.train(
        batch_size = hparams['batch_size'],
        learning_rate = hparams['learning_rate']
    )
    accuracy = model.test()

    # send final accuracy to NNI
    nni.report_final_result(accuracy)
```

And finally, we define the main entry point for the trial:

```
if __name__ == '__main__':

    # Manual HyperParameters
    hparams = {
        'filter_size':   32,
        'kernel_size':   3,
        'l1_size':       64,
        'batch_size':    512,
        'learning_rate': 1e-3,
    }

    # NNI HyperParameters
    # Run safely without NNI Experiment Context
    nni_hparams = nni.get_next_parameter()
    hparams.update(nni_hparams)

    trial(hparams)
```

Remember that a trial script can be executed in stand-alone mode, so you can run a ch2/lenet_hpo/tf_trial.py to test its execution with custom parameters.

PyTorch LeNet Implementation

In this section, we will study the LeNet model's implementation using PyTorch. Listing 2-11 demonstrates the implementation of the PyTorch LeNet model for the hyperparameter optimization.

We import necessary modules:

Listing 2-11. LeNet. PyTorch implementation. ch2/lenet_hpo/pt_lenet_model.py

```
import numpy as np
import nni
import torch
import torch.nn as nn
import torch.nn.functional as F
import torch.optim as optim
from sklearn.metrics import accuracy_score
from ch2.utils.datasets import mnist_dataset
```

Next, we define the LeNet model with three layer hyperparameters:

```
class PtLeNetModel(nn.Module):

    def __init__(self, filter_size, kernel_size, l1_size):
        super(PtLeNetModel, self).__init__()
```

This implementation will use lazy layer initialization, so we explicitly save the `l1_size` hyperparameter:

```
        # Saving l1_size HyperParameter
        self.l1_size = l1_size
```

After that, we initialize convolutional layers:

```
        self.conv1 = nn.Conv2d(
            in_channels = 1,
            out_channels = filter_size,
            kernel_size = kernel_size
        )
```

```
        self.conv2 = nn.Conv2d(
            in_channels = filter_size,
            out_channels = filter_size * 2,
            kernel_size = kernel_size
        )
```

We don't initialize the first linear layer, but we use lazy initialization. To initialize a linear layer, we must specify an in_features value. But this is not so simple. We need to know the dimension of the tensor, which the previous layers will produce. To do this, sometimes, you have to do complex calculations. It is easier to calculate the dimension of this tensor at the first call and at this moment initialize the linear layer.

```
        # Lazy fc1 Layer Initialization
        self.fc1__in_features = 0
        self._fc1 = None
        self.fc2 = nn.Linear(l1_size, 10)
```

Lazy layer initialization:

```
    @property
    def fc1(self):
        if self._fc1 is None:
            self._fc1 = nn.Linear(
                self.fc1__in_features,
                self.l1_size
            )
        return self._fc1
```

LeNet is a straightforward model which passes calculation results from one layer to another:

```
    def forward(self, x):
        x = torch.sigmoid(self.conv1(x))
        x = F.max_pool2d(x, 2, 2)
        x = torch.sigmoid(self.conv2(x))
        x = F.max_pool2d(x, 2, 2)
```

```
        # Flatting all dimensions but batch-dimension
        if not self.fc1__in_features:
            self.fc1__in_features = np.prod(x.shape[1:])

        x = x.view(-1, self.fc1__in_features)

        # FC1 initializes lazy
        x = torch.sigmoid(self.fc1(x))
        x = self.fc2(x)
        return F.log_softmax(x, dim = 1)
```

The training method uses two training hyperparameters: `batch_size` and `learning rate`:

```
    def train_model(self, learning_rate, batch_size):
```

We prepare training dataset:

```
        (x, y), _ = mnist_dataset()
        x = torch.from_numpy(x).float()
        y = torch.from_numpy(y).long()

        # Permute dimensions for PyTorch Convolutions
        x = torch.permute(x, (0, 3, 1, 2))
        dataset_size = x.shape[0]
```

Initialize Adam optimizer:

```
        optimizer = optim.Adam(
            self.parameters(),
            lr = learning_rate
        )
```

Vanilla PyTorch does not have built-in batch training. Therefore, we manually split the dataset into batches and perform epoch loop and batch loop:

```
        self.train()
        for epoch in range(1, 10 + 1):
            # Random permutations for batch training
            permutation = torch.randperm(dataset_size)
            for bi in range(1, dataset_size, batch_size):
```

```
            # Creating New Batch
            indices = permutation[bi:bi + batch_size]
            batch_x, batch_y = x[indices], y[indices]
```

Model parameter optimization using cross-entropy loss function:

```
            optimizer.zero_grad()
            output = self(batch_x)
            loss = F.cross_entropy(output, batch_y)
            loss.backward()
            optimizer.step()
```

At the end of each epoch, we calculate the model accuracy and return it to NNI as an intermediate result:

```
        output = self(x)
        predict = output.argmax(dim = 1, keepdim = True)
        accuracy = round(accuracy_score(predict, y), 4)
        print(F'Epoch: {epoch}| Accuracy: {accuracy}')
        # report intermediate result
        nni.report_intermediate_result(accuracy)
```

And the last method we need to define is model testing. We load the test MNIST dataset and perform the classification by measuring its accuracy:

```
    def test_model(self):
        self.eval()
        # Preparing Test Dataset
        _, (x, y) = mnist_dataset()
        x = torch.from_numpy(x).float()
        y = torch.from_numpy(y).long()
        x = torch.permute(x, (0, 3, 1, 2))

        with torch.no_grad():
            output = self(x)
            predict = output.argmax(dim = 1, keepdim = True)
            accuracy = round(accuracy_score(predict, y), 4)

        return accuracy
```

Well, since the implementation of PyTorch LeNet model is ready, we can implement the NNI trial script using Listing 2-12.

We import necessary modules and pass code root directory to system path:

Listing 2-12. NNI trial script with TensorFlow LeNet implementation. ch2/lenet_hpo/pt_trial.py

```
import os
import sys
import nni

# For NNI use relative import for user-defined modules
SCRIPT_DIR = os.path.dirname(os.path.abspath(__file__)) + '/../..'
sys.path.append(SCRIPT_DIR)

from ch2.lenet_hpo.pt_lenet_model import PtLeNetModel
```

The `trial` method initializes the model, trains it, tests it, and returns the NNI metric:

```
def trial(hparams):
    model = PtLeNetModel(
        filter_size = hparams['filter_size'],
        kernel_size = hparams['kernel_size'],
        l1_size = hparams['l1_size']
    )
    model.train_model(
        batch_size = hparams['batch_size'],
        learning_rate = hparams['learning_rate']
    )
    accuracy = model.test_model()
    nni.report_final_result(accuracy)
```

And finally, we define the main entry point for the trial:

```
if __name__ == '__main__':
    # Manual HyperParameters
    hparams = {
        'filter_size':   32,
        'kernel_size':   3,  #5,
```

```
        'l1_size':        64, #1024,
        'batch_size':     512, #32,
        'learning_rate': 1e-2, #1e-4,
    }

    # NNI HyperParameters
    # Run safely without NNI Experiment Context
    nni_hparams = nni.get_next_parameter()
    hparams.update(nni_hparams)

    trial(hparams)
```

Remember that a trial script can be executed in stand-alone mode, so you can run a ch2/lenet_hpo/pt_trial.py to test its execution with custom parameters.

Performing LeNet HPO Experiment

And so now, we are all set for our first HPO study. Any real-world experiment can take several hours to several weeks, which is natural because each trial creates a unique model and training method. And the process of training can take quite a long time, depending on a model design and dataset. Some of the experiments in this book took quite a long time. Therefore, you can skip a complete experiment run or limit the number of trials with the `maxTrialNumber` setting. Keep in mind that if you run a limited experiment, your results may differ significantly from those presented in the book. For each experiment, I will give the time it took to complete it on a specific machine. We need to configure the NNI experiment and run it. Listing 2-13 contains the configuration for the LeNet HPO Experiment.

Listing 2-13. LeNet HPO Experiment configuration. ch2/lenet_hpo/config.yml

```
trialConcurrency: 4
searchSpaceFile: search_space.json
trialCodeDirectory: .
```

Uncomment PyTorch trial line to run the experiment using PyTorch implementation:

```
trialCommand: python3 tf_trial.py
#trialCommand: python3 pt_trial.py
```

The search space contains $3^5 = 243$ elements. This is a small search space, and we can use the Grid Search Tuner here:

```
tuner:
  name: GridSearch
trainingService:
  platform: local
```

The experiment can be run as follows:

```
nnictl create --config ch2/lenet_hpo/config.yml
```

Note Duration ~ 2 hours on Intel Core i7 with CUDA (GeForce GTX 1050)

The experiment returned the following best trial hyperparameters:

- **learning_rate**: 0.001
- **batch_size**: 256
- **l1_size**: 64
- **kernel_size**: 5
- **filter_size**: 32

The best trial showed a **0.9885** result. And this is an acceptable outcome. We can assume that LeNet supplied by best hyperparameters recognizes handwritten numbers with 98.85% accuracy on test dataset.

In Figure 2-12, we can observe the top 1% of the top trials in the hyperparameters panel.

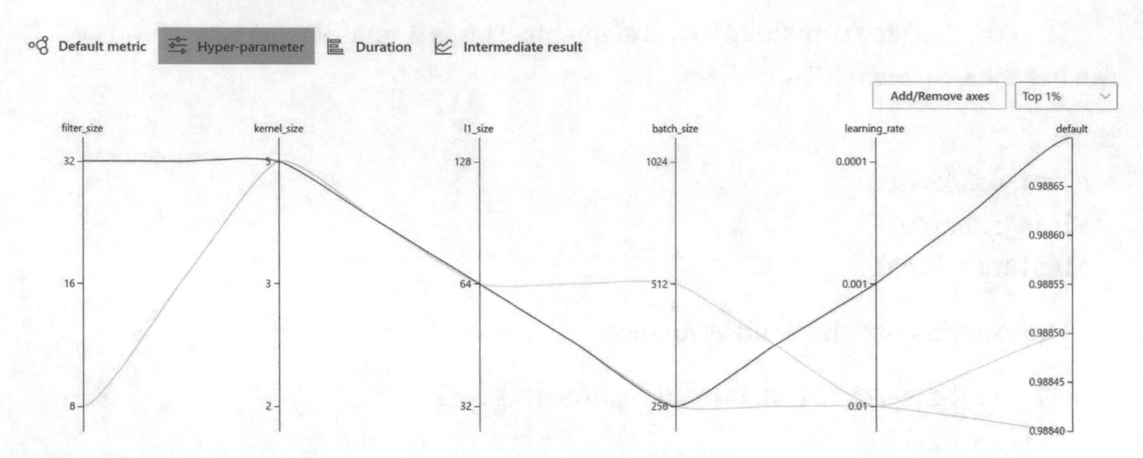

***Figure 2-12.** Hyperparameter panel of LeNet HPO top 1% trials*

Figure 2-12 demonstrates that all the best results have `kernel_size` = 5. Otherwise, the best results have no dependencies among its hyperparameters.

After completing a study, I like to visualize the results. We already have accuracy metric. But it will still be interesting to glance at the images that the LeNet model could not classify correctly. Perhaps the achieved accuracy of 98.85% is the best possible accuracy? Maybe the test dataset contains samples that cannot be correctly classified? Listing 2-14 displays the first nine failed predictions.

We import necessary modules:

Listing 2-14. LeNet failed predictions. ch2/lenet_hpo/display_mnist_failed_predictions.py

```
from math import floor
import numpy as np
from ch2.lenet_hpo.tf_lenet_model import TfLeNetModel
import matplotlib.pyplot as plt
import tensorflow as tf
from ch2.utils.datasets import mnist_dataset
```

LeNet model is initialized using best layer hyperparameters:

```
# Best Hyperparameters
hparams = {
    "learning_rate": 0.001,
    "batch_size":    256,
```

```
    "l1_size":       64,
    "kernel_size":   5,
    "filter_size":   32
}

# Making this script Reproducible
tf.random.set_seed(1)

# Initializing LeNet Model
model = TfLeNetModel(
    filter_size = hparams['filter_size'],
    kernel_size = hparams['kernel_size'],
    l1_size = hparams['l1_size']
)
```

And after that, we train the model using the best training hyperparameters:

```
# Model Training
model.train(
    batch_size = hparams['batch_size'],
    learning_rate = hparams['learning_rate']
)
```

Trained model makes its predictions on test MNIST dataset:

```
# MNIST Dataset
(_, _), (x_test, y_test) = mnist_dataset()

# Predictions
output = model(x_test)
y_pred = tf.argmax(output, 1)
```

Collecting the first nine failed predictions:

```
number_of_fails_left = 9
fails = []
for i in range(len(x_test)):
    if number_of_fails_left == 0:
        break
```

```
    if y_pred[i] != y_test[i]:
        fails.append((x_test[i], (y_pred[i], y_test[i])))
        number_of_fails_left -= 1
```

Displaying failed predictions:

```
fig, axs = plt.subplots(3, 3)
for i in range(len(fails)):
    sample, (pred, actual) = fails[i]
    img = np.array(sample, dtype = 'float')
    img = img * 255
    pixels = img.reshape((28, 28))
    ax = axs[floor(i / 3), i % 3]
    ax.set_title(f'#{i+1}: {actual} ({pred})')
    ax.set_xticks([])
    ax.set_yticks([])
    ax.imshow(pixels, cmap = 'gray')
plt.show()
```

Figure 2-13 displays LeNet failed predictions. To be honest, the samples #1, #3, #4, #5, #6, and #8 are really difficult to classify. I don't think that a reader would recognize the number 2 in sample #4. Therefore, we do not need to demand 100% accuracy from our model. But still, I think there is room for improvement.

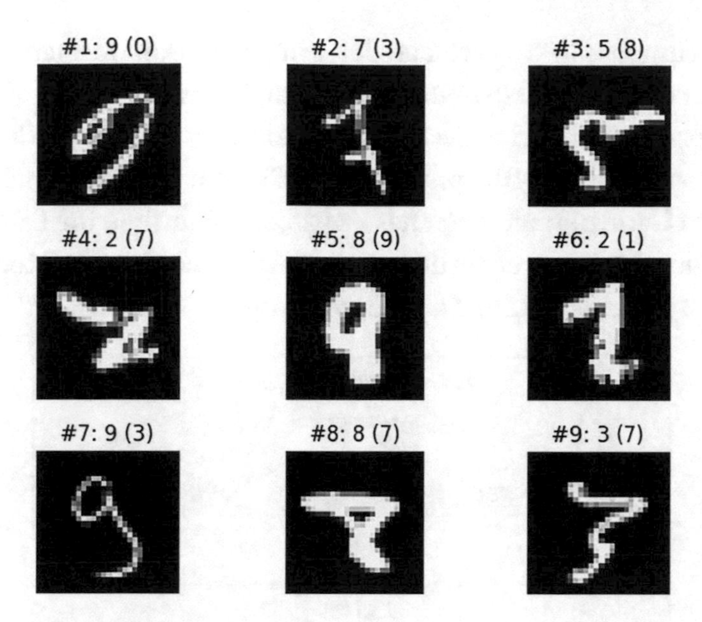

Figure 2-13. *LeNet failed predictions*

Congratulations! We have completed our first real-world study. We defined the problem, chose the model frame (LeNet), expressed the search space, made model implementations, and found the best hyperparameters for the problem. But this is only the beginning. Let's continue the research and see what other exciting results we can achieve.

Upgrading LeNet with ReLU and Dropout

Libraries: TensorFlow (Keras API), PyTorch

Experienced data scientists may have the question: Hey?! Why didn't we use the dropout layer and rectified linear unit (ReLU) as the activation function for LeNet model in the previous section? Because the original LeNet model did not use the ReLU activation. And the dropout technique was first introduced in 2012, 22 years after creating the LeNet architecture. One of the main problems with the LeNet concept was that it used sigmoid as an activation function. The sigmoid activation function led to slower training and a vanishing gradient problem. The dropout technique is also one of the most popular regularization methods. Nowadays, we cannot imagine a successful pattern recognition model that does not use ReLU and dropout layers. Let's pretend that we have never heard anything about ReLU and dropout, and someone advises us to inject these techniques into a LeNet model to improve its performance. We can do research using HPO that will help us find the best architecture.

Let's introduce an `activation` design hyperparameter responsible for choosing the activation function. To simplify the problem, we will use a one-for-all policy. This means that if the `activation` hyperparameter has a `sigmoid` value, then the LeNet model will have a sigmoid function for all activations. The same is true if `activation` has a `relu` value. Figure 2-14 presents `activation` hyperparameter.

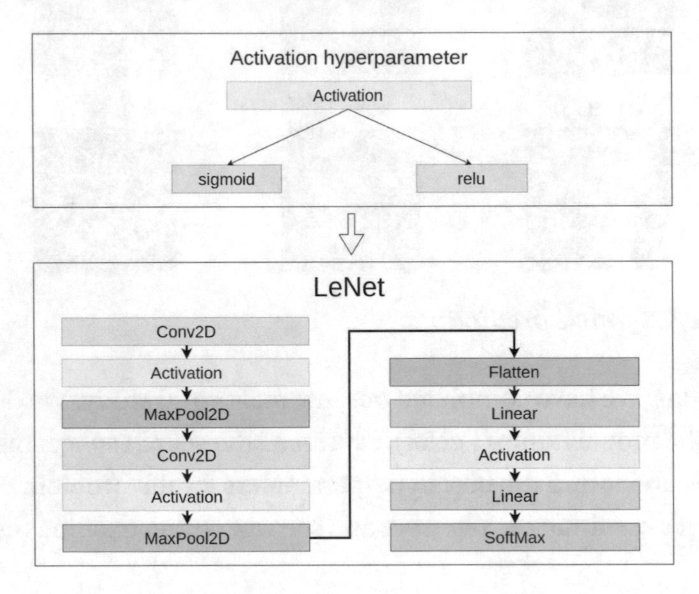

Figure 2-14. *Activation design hyperparameter*

Typically, a dropout layer is inserted between linear layers. But we initially do not know whether this technique would be effective, so we have to make possible two variants of LeNet architecture: with dropout layer and without dropout layer. To do this, we use the `use_dropout` design hyperparameter. If `use_dropout` is `0`, then the LeNet model does not use the dropout layer, and if `use_dropout` is `1`, then the LeNet model uses the dropout layer. At the same time, the dropout layer will be tested using three different p (dropout rate) values: 0.3, 0.5, and 0.7. Figure 2-15 presents `use_dropout` design hyperparameter.

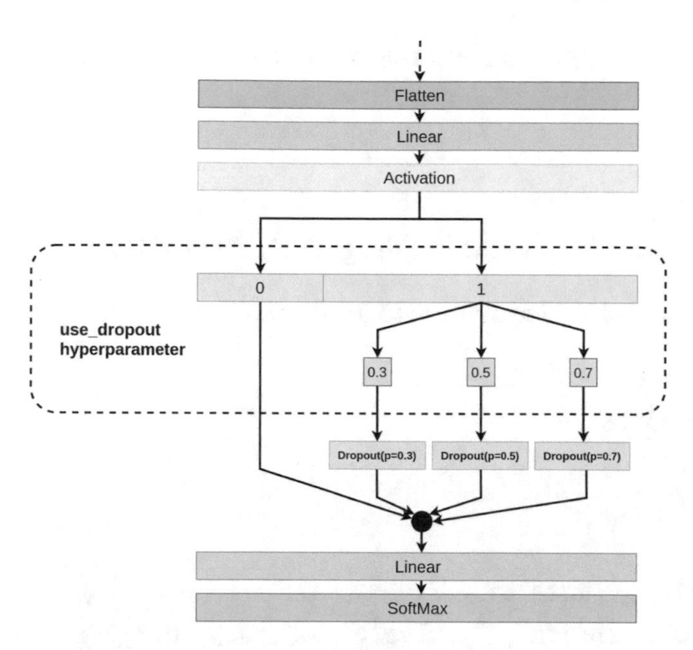

Figure 2-15. *Dropout design hyperparameter*

Each model design works well with specific layer hyperparameters. Therefore, we also need to include layer hyperparameters in the search space. In this experiment, we will use the same hyperparameters we used in the previous section. But we will choose slightly different values for them:

- `filter_size:` `16, 32`
- `kernel_size:` `5, 7`
- `l1_size:` `64, 128, 256`
- `batch_size:` `512, 1024`
- `learning_rate:` `0.001, 0.0001`

Listing 2-15 presents hyperparameter constraints we made earlier as NNI search space.

Listing 2-15. LeNet Upgrade HPO search space. ch2/lenet_upgrade/search_space.json

```
{
  "activation": {
    "_type": "choice", "_value": ["sigmoid", "relu"]},
```

Here, we use nested choice method to implement `use_dropout` hyperparameter:

```
  "use_dropout": {
    "_type": "choice",
    "_value": [
      {"_name": 0},
      {
        "_name": 1, "rate":
        {"_type": "choice", "_value": [0.3, 0.5, 0.7]}
      }
    ]
  },
  "filter_size": {
    "_type": "choice", "_value": [16, 32]},
  "kernel_size": {
    "_type": "choice", "_value": [5, 7]},
  "l1_size": {
    "_type": "choice", "_value": [64, 128, 256]},
  "batch_size": {
    "_type": "choice", "_value": [512, 1024]},
  "learning_rate": {
    "_type": "choice", "_value": [0.001, 0.0001]}
}
```

And just like in the previous section, the next step is to make TensorFlow and PyTorch implementations of the LeNet Upgrade model.

TensorFlow LeNet Upgrade Implementation

In this section, we will examine the LeNet Upgrade model's implementation using TensorFlow (Keras API).

We will examine only `__init__` and `call` methods in `TfLeNetUpgradeModel`. Other methods are the same as in `TfLeNetModel` (ch2/lenet_hpo/tf_lenet_model.py). Listing 2-16 shows the LeNet Upgrade model with six hyperparameters.

Listing 2-16. LeNet Upgrade. TensorFlow implementation. ch2/lenet_upgrade/tf_lenet_upgrade_model.py

```
class TfLeNetUpgradeModel(Model):

    def __init__(
            self,
            filter_size,
            kernel_size,
            l1_size,
            activation,
            use_dropout,
            dropout_rate = None
    ):
        super().__init__()
```

First layer stack with `activation` variable:

```
        self.conv1 = Conv2D(
            filters = filter_size,
            kernel_size = kernel_size,
            activation = activation
        )
        self.pool1 = MaxPool2D(pool_size = 2)
        self.conv2 = Conv2D(
            filters = filter_size * 2,
            kernel_size = kernel_size,
            activation = activation
        )
```

```
        self.pool2 = MaxPool2D(pool_size = 2)
        self.flatten = Flatten()
        self.fc1 = Dense(
            units = l1_size,
            activation = activation
        )
```

We add dropout layer if use_dropout and identity layer otherwise:

```
        if use_dropout:
            self.drop = Dropout(rate = dropout_rate)
        else:
            self.drop = tf.identity
```

Final linear layer stack:

```
        self.fc2 = Dense(
            units = 10,
            activation = 'softmax'
        )
```

LeNet Upgrade model invokes each layer sequentially:

```
    def call(self, x, **kwargs):
        x = self.conv1(x)
        x = self.pool1(x)
        x = self.conv2(x)
        x = self.pool2(x)
        x = self.flatten(x)
        x = self.fc1(x)
        x = self.drop(x)
        return self.fc2(x)
```

And after implementing LeNetUpgradeModel, we can implement the NNI trial script using Listing 2-17.

We import necessary modules and pass code root directory to system path:

Listing 2-17. NNI trial script with TensorFlow LeNetUpgrade implementation. ch2/lenet_upgrade/tf_trial.py

```
import os
import sys
import nni

# We use relative import for user-defined modules
SCRIPT_DIR = os.path.dirname(os.path.abspath(__file__)) + '/../..'
sys.path.append(SCRIPT_DIR)

from ch2.lenet_upgrade.tf_lenet_upgrade_model import TfLeNetUpgradeModel
```

The trial method initializes the model, trains it, tests it, and returns the NNI metric:

```
def trial(hparams):
    use_dropout = bool(hparams['use_dropout']['_name'])
    model_params = {
        "filter_size": hparams['filter_size'],
        "kernel_size":   hparams['kernel_size'],
        "l1_size":     hparams['l1_size'],
        "activation":  hparams['activation'],
        "use_dropout": use_dropout
    }
    if use_dropout:
        model_params['dropout_rate'] = hparams['use_dropout']['rate']
    model = TfLeNetUpgradeModel(**model_params)
    model.train(
        batch_size = hparams['batch_size'],
        learning_rate = hparams['learning_rate']
    )
    accuracy = model.test()

    # send final accuracy to NNI
    nni.report_final_result(accuracy)
```

Next, we define the main entry point for the trial:

```
if __name__ == '__main__':

    # Manual HyperParameters
    hparams = {
        'use_dropout':   {'_name': 1, 'rate': 0.5},
        'activation':    'relu',
        'filter_size':   32,
        'kernel_size':   3,
        'l1_size':       64,
        'batch_size':    512,
        'learning_rate': 1e-3,
    }

    # NNI HyperParameters
    # Run safely without NNI Experiment Context
    nni_hparams = nni.get_next_parameter()
    hparams.update(nni_hparams)

    trial(hparams)
```

Remember that a trial script can be executed in stand-alone mode, so you can run a ch2/lenet_upgrade/tf_trial.py to test its execution with custom parameters.

PyTorch LeNet Upgrade Implementation

In this section, we will examine the LeNet Upgrade model's implementation using PyTorch.

We will examine only `__init__` and `forward` methods in `PtLeNetUpgradeModel`. Other methods are the same as in `PtLeNetModel` (ch2/lenet_hpo/pt_lenet_model.py). Listing 2-18 shows the LeNet Upgrade model with six hyperparameters:

Listing 2-18. LeNet Upgrade. PyTorch implementation. ch2/lenet_upgrade/pt_lenet_upgrade_model.py

```
class PtLeNetUpgradeModel(nn.Module):

    def __init__(
            self,
            filter_size,
            kernel_size,
            l1_size,
            activation,
            use_dropout,
            dropout_rate = None
    ):
        super(PtLeNetUpgradeModel, self).__init__()
```

We set `self.act` layer by activation variable:

```
        # Activation Function
        if activation == 'relu':
            self.act = torch.relu
        elif activation == 'sigmoid':
            self.act = torch.sigmoid
        else:
            raise Exception(f'Unknown activation: {activation}')
```

We add dropout layer if `use_dropout` and identity layer otherwise:

```
        if use_dropout:
            self.drop = nn.Dropout(p = dropout_rate)
        else:
            self.drop = nn.Identity()
```

Next, we set other LeNet layers:

```
        # Saving l1_size HyperParameter
        self.l1_size = l1_size
```

```
        self.conv1 = nn.Conv2d(
            1,
            filter_size,
            kernel_size = kernel_size
        )
        self.conv2 = nn.Conv2d(
            filter_size,
            filter_size * 2,
            kernel_size = kernel_size
        )
        # Lazy fc1 Layer Initialization
        self.fc1__in_features = 0
        self._fc1 = None
        self.fc2 = nn.Linear(l1_size, 10)
```

LeNet Upgrade model invokes each layer sequentially:

```
    def forward(self, x):
        x = self.act(self.conv1(x))
        x = F.max_pool2d(x, 2, 2)
        x = self.act(self.conv2(x))
        x = F.max_pool2d(x, 2, 2)
        # Flatting all dimensions but batch-dimension
        self.fc1__in_features = np.prod(x.shape[1:])
        x = x.view(-1, self.fc1__in_features)
        x = self.act(self.fc1(x))
        x = self.drop(x)
        x = self.fc2(x)
        return F.log_softmax(x, dim = 1)
```

And after implementing `LeNetUpgradeModel`, we can implement the NNI trial script using Listing 2-19.

We import necessary modules and pass code root directory to system path:

Listing 2-19. NNI trial script with PyTorch LeNetUpgrade implementation. ch2/lenet_upgrade/pt_trial.py

```
import os
import sys
import nni

# We use relative import for user-defined modules
SCRIPT_DIR = os.path.dirname(os.path.abspath(__file__)) + '/../..'
sys.path.append(SCRIPT_DIR)

from ch2.lenet_upgrade.pt_lenet_upgrade_model import PtLeNetUpgradeModel
```

The `trial` method initializes the model, trains it, tests it, and returns the NNI metric:

```
def trial(hparams):
    use_dropout = bool(hparams['dropout']['_name'])
    model_params = {
        "filter_size": hparams['filter_size'],
        "kernel_size": hparams['kernel_size'],
        "l1_size":     hparams['l1_size'],
        "activation":  hparams['activation'],
        "use_dropout": use_dropout
    }
    if use_dropout:
        model_params['dropout_rate'] = hparams['dropout']['rate']

    model = PtLeNetUpgradeModel(**model_params)

    model.train_model(
        batch_size = hparams['batch_size'],
        learning_rate = hparams['learning_rate']
    )
    accuracy = model.test_model()
    nni.report_final_result(accuracy)
```

Next, we define the main entry point for the trial:

```
if __name__ == '__main__':
    # Manual HyperParameters
    hparams = {
        'dropout':       {'_name': 1, 'rate': 0.5},
        'activation':    'relu',
        'filter_size':   32,
        'kernel_size':   3,
        'l1_size':       64,
        'batch_size':    512,
        'learning_rate': 1e-3,
    }

    # NNI HyperParameters
    # Run safely without NNI Experiment Context
    nni_hparams = nni.get_next_parameter()
    hparams.update(nni_hparams)

    trial(hparams)
```

Remember that a trial script can be executed in stand-alone mode, so you can run a ch2/lenet_upgrade/pt_trial.py to test its execution with custom parameters.

Performing LeNet Upgrade HPO Experiment

We are now ready to run our second experiment. We can consider this experiment as a battle between Vanilla LeNet model and LeNet enhanced by ReLU and dropout. Please be aware that the results of this study will be specific to the MNIST problem. An experiment launched on a different dataset can lead to completely different results. Listing 2-20 defines the configuration of the LeNet Upgrade experiment.

Listing 2-20. LeNet Upgrade HPO Experiment configuration. ch2/lenet_upgrade/config.yml

```
trialConcurrency: 2
```

We will limit the number of trials:

```
maxTrialNumber: 300
searchSpaceFile: search_space.json
trialCodeDirectory: .
```

Uncomment PyTorch trial line to run the experiment using PyTorch implementation:

```
trialCommand: python3 tf_trial.py
#trialCommand: python3 pt_trial.py
```

GridSearch Tuner cannot be used for search spaces that utilize nested choice, so we pick Random Search Tuner:

```
tuner:
  name: Random
trainingService:
  platform: local
```

The experiment can be run as follows:

```
nnictl create --config ch2/lenet_upgrade/config.yml
```

Note Duration ~ 3 hours on Intel Core i7 with CUDA (GeForce GTX 1050)

The experiment returned the following best trial hyperparameters:

- **activation**: relu
- **use_dropout**: 1
    - **rate**: 0.5
- **filter_size**: 32
- **kernel_size**: 5
- **l1_size**: 256
- **batch_size**: 512
- **learning_rate**: 0.001

The best trial demonstrated a **0.9923** result, which is a significant improvement over the 0.9885 we got in the previous section. We see that the best hyperparameter combination uses ReLU activation and dropout (p=0.5) layer. Could this mean that LeNet empowered by ReLU and dropout won this battle? Let's look at the top 1% trials in the hyperparameter panel (Figure 2-16) to answer this question.

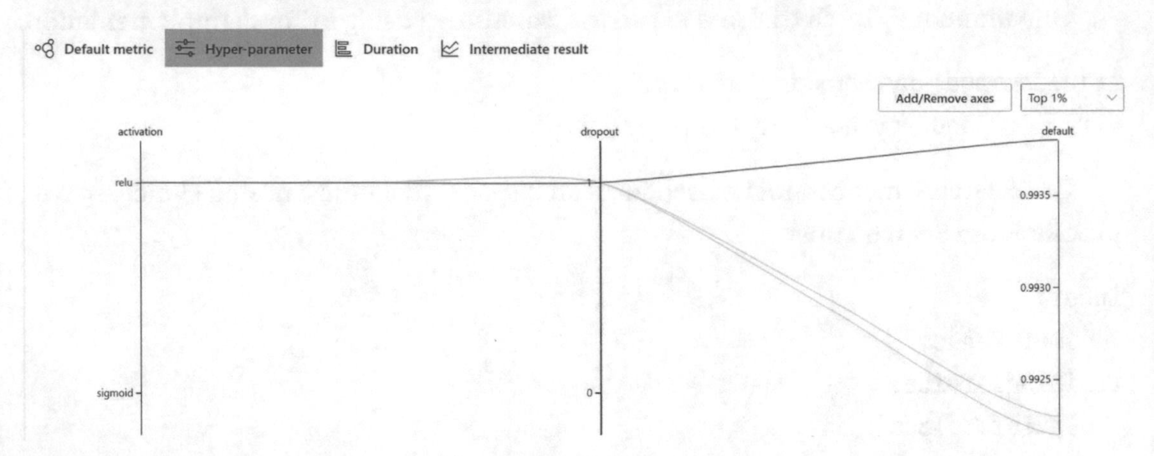

Figure 2-16. *Hyperparameter panel of LeNet Upgrade HPO top 1% trials*

Figure 2-16 demonstrates that all three best hyperparameter combinations have the following hyperparameter values:

- **`activation`**: relu
- **`use_dropout`**: 1

which can be considered as solid evidence in favor of using the ReLU and dropout techniques. Of course, this result may seem somewhat obvious, but this is only because you already know about the benefits of using ReLU and dropout layers. Initially, this fact did not seem so obvious and required practical evidence, which we have just demonstrated.

Finally, let's take a look at the MNIST database samples that the upgraded LeNet model failed to classify. These samples are presented in Figure 2-17.

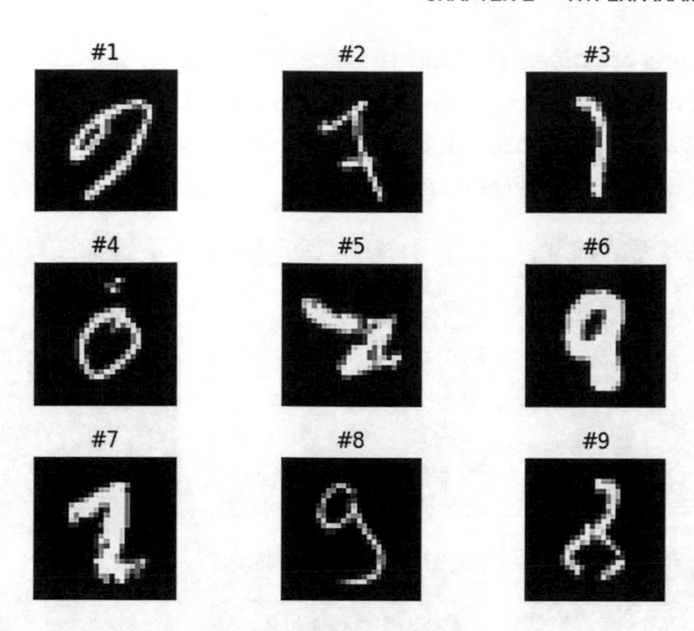

Figure 2-17. *Upgraded LeNet model fails*

I deliberately didn't print the correct results in Figure 2-17. Take one minute, write down your answer for each sample, and compare it with the correct answers in the following:

Answers. #1: 9. #2: 7. #3: 7. #4: 0. #5: 2. #6: 8. #7: 2. #8: 9. #9. 8

If you haven't made a single mistake, then I admire you! I guessed only four numbers. If a human has difficulties recognizing some handwritten characters, then the neural network is already close to its performance threshold. The current result of the upgraded LeNet model we have developed is close to the best.

This section demonstrates how new deep learning techniques can be injected into existing architecture. We defined a search space to choose the best design combination. HPO chose an upgraded model design that significantly improved the performance of the original model. And this is a very simple and useful technique that will allow you to uptune your models using the latest advances in machine learning.

From LeNet to AlexNet

Libraries: TensorFlow (Keras API), PyTorch (PyTorch Lightning)

Well, handwriting recognition is a rather important task, but it seems that it's time to move on to more complicated problems. Let's try to classify more complex objects. How about developing a model that will classify humans and horses? Figure 2-18 shows samples of "humans or horses" dataset. This dataset contains 300×300 color images, that is, (300, 300, 3) tensors. Obviously, the image of a human or a horse is more complex than a 28×28 grayscale image of a handwritten number. And perhaps, we will need to evolve the LeNet model that we considered earlier. We will call it *LeNet Evolution* model.

Figure 2-18. *Humans vs. horses dataset*

Let's look at the architecture of the LeNet model again. LeNet model design can be divided into two components: *feature extraction* and *decision making*. Indeed, the convolution layer stack (*Conv2D → Activation → MaxPool2D → Conv2D → Activation → MaxPool2D → Flatten*) is responsible for extracting image patterns, that is, feature extraction. At the same time, the fully connected layer stack (*Linear → Activation → Linear → SoftMax*) is responsible for selecting particular patterns to classify an input object. Figure 2-19 shows the areas of responsibility for each component.

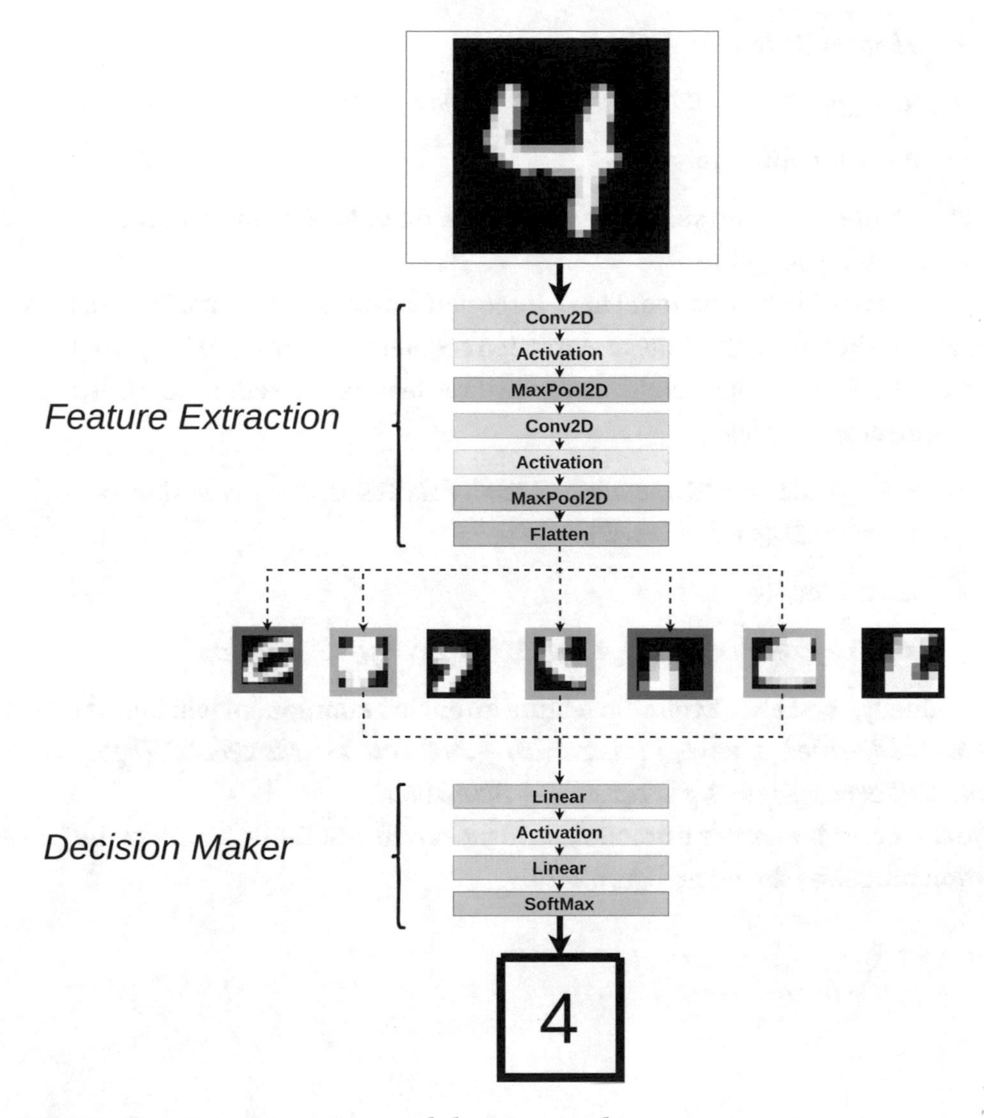

Figure 2-19. *Feature extraction and decision maker components*

Since human and horse images are more complex, we need to make the feature extraction component more sophisticated. Two types of layer sequences are usually responsible for extracting image patterns: *Conv2D → Activation → MaxPool2D* and *Conv2D → Activation*. We can build an experiment that will inject different feature extraction sequences to find the best model design to solve the "human and horses" classification problem. In the following, we define three types of feature extraction layer sequences adding none as an empty sequence:

- **simple**: *Conv2D → Activation*
- **with_pool**: *Conv2D → Activation → MaxPool2D*
- **none**: Identity layer

Each feature extraction sequence will have additional layer hyperparameters: filters, kernel, pool_size.

Let the LeNet Evolution model have three pattern extraction slots. Each of these slots can be filled with one of the feature extraction sequences: simple, with_pool, none. For example, a LeNet Evolution model can have three feature extraction slots filled with the following layer sequences:

- **with_pool**: *Conv2D(kernel_size=5, filters=16) → Activation → MaxPool2D(pool_size=3)*
- **none**: Identity
- **simple**: *Conv2D(kernel_size=3, filters=8)→ Activation*

And finally, the LeNet Evolution feature extraction component will look the following way: *Conv2D(kernel_size=5, filters=16) → Activation → MaxPool2D(pool_size=3) → Conv2D(kernel_size=3, filters=8) → Activation.*

We can consider feature extraction sequences as the building blocks of the LeNet Evolution model, as shown in Figure 2-20.

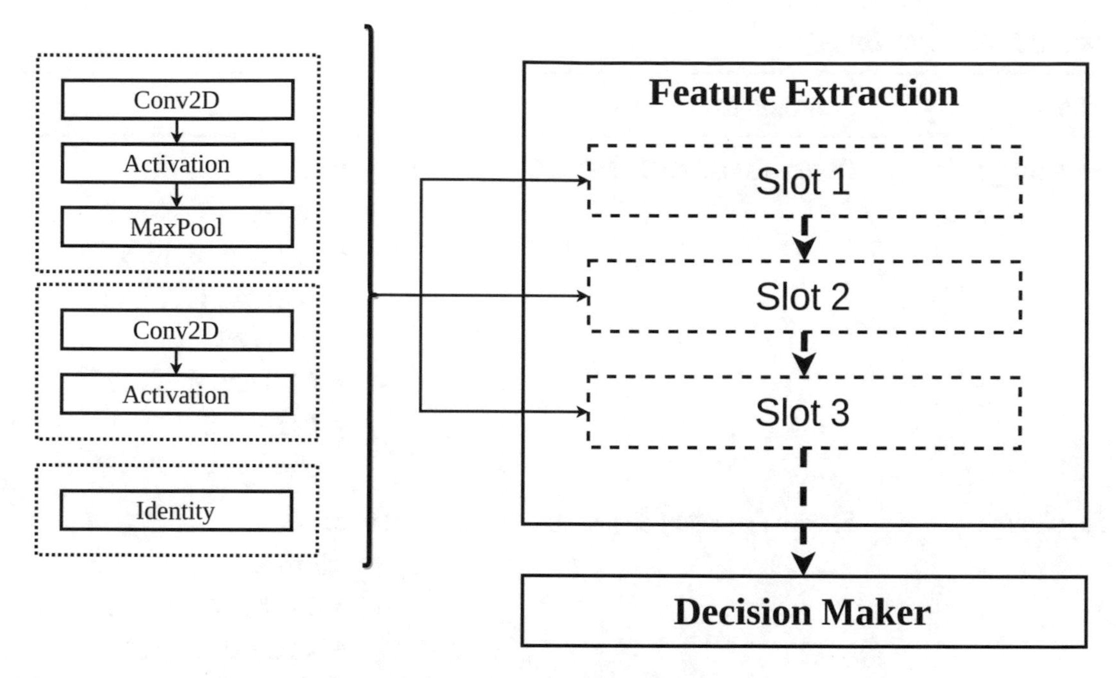

***Figure 2-20.** LeNet Evolution feature extraction component*

And so we can define the three design hyperparameters of the LeNet Evolution model. Table 2-4 provides LeNet Evolution model design hyperparameters.

***Table 2-4.** LeNet Evolution feature extraction hyperparameters*

Name	Description	Values

fe_slot_1	Feature Extraction Sequence 1	• none
		• simple
		  • filters: 8, 16, 32
		  • kernel: 5, 7, 9, 11
		• with_pool:
		  • filters: 8, 16, 32
		  • kernel: 5, 7, 9, 11
		  • pool_size: 3, 5, 7

(*continued*)

Table 2-4. *(continued)*

Name	Description	Values

fe_slot_2	Feature Extraction Sequence 2	• none
		• simple
		  • filters: 8, 16, 32
		  • kernel: 5, 7, 9, 11
		• with_pool:
		  • filters: 8, 16, 32
		  • kernel: 5, 7, 9, 11
		  • pool_size: 3, 5, 7

fe_slot_3	Feature Extraction Sequence 3	• none
		• simple
		  • filters: 8, 16, 32
		  • kernel: 5, 7, 9, 11
		• with_pool:
		  • filters: 8, 16, 32
		  • kernel: 5, 7, 9, 11
		  • pool_size: 3, 5, 7

Since the feature extraction component of the LeNet Evolution model returns more features than it did with the MNIST problem, we should also let the experiment create a more advanced decision maker component. The decision maker component can be improved by adding an extra linear layer. This is the easiest and most efficient way to enhance a decision maker component. Since we proved the sustainability of the dropout layer and ReLU activation in the previous section, they will also be used in the decision maker component. Figure 2-21 demonstrates two variants of the decision maker component.

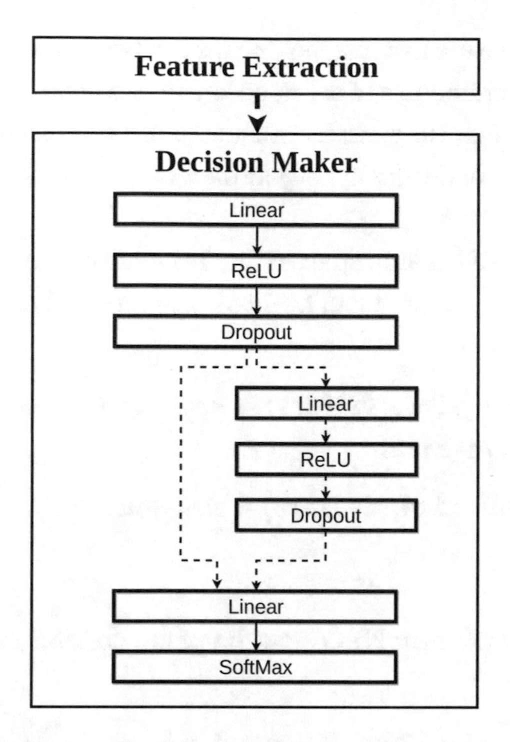

Figure 2-21. *LeNet Evolution decision maker component*

The design of the decision maker component will be determined by the following hyperparameters shown in Table 2-5.

Table 2-5. *LeNet Evolution decision maker hyperparameters*

Name	Description	Values
`l1_size`	Output size of first linear layer	512, 1024, 2048
`l2_size`	Output size of second linear layer	0, 512, 1024 If 0 value is chosen, then the second linear layer is skipped
`dropout_rate`	Dropout probability of dropout layers	0.3, 0.5, 0.7

Also, we will use `learning_rate` as a training hyperparameter with the following options: 0.001, 0.0001.

In this experiment, we are not just looking for the best hyperparameters, but we are trying to create a new architecture of the deep learning model based on the principles of the original LeNet model. Here, we try not just to tune an existing model but also to create a new one. The list of design hyperparameters is responsible for unique deep learning model design.

Listing 2-21 defines NNI search space of the LeNet Evolution model.

The first feature slot `fe_slot_1` can be filled with one of these feature extraction sequences:

- *Conv2D*(`kernel_size, filters`) → *Activation* → *MaxPool2D*(`pool_size`)
- *Conv2D*(`kernel_size, filters`) → *Activation*
- *None*

Listing 2-21. LeNet Evolution HPO search space. ch2/lenet_to_alexnet/search_space.json

```
{
  "fe_slot_1": {
    "_type": "choice",
    "_value": [
      {"_name": "none"},
      {
        "_name": "simple",
        "filters": {"_type": "choice", "_value": [8, 16, 32]},
        "kernel": {"_type": "choice", "_value": [5, 7, 9, 11]}
      },
      {
        "_name": "with_pool",
        "filters": {"_type": "choice", "_value": [8, 16, 32]},
        "kernel": {"_type": "choice", "_value": [5, 7, 9, 11]},
        "pool_size": {"_type": "choice", "_value": [3, 5, 7]}
      }
    ]
  },
```

The second and third feature extraction slots (fe_slot_2, fe_slot_3) have the same values set as fe_slot_1:

```
"fe_slot_2": {
  "_type": "choice",
  "_value": [
    {"_name": "none"},
    {
      "_name": "simple",
      "filters": {"_type": "choice", "_value": [8, 16, 32]},
      "kernel": {"_type": "choice", "_value": [5, 7, 9, 11]}
    },
    {
      "_name": "with_pool",
      "filters": {"_type": "choice", "_value": [8, 16, 32]},
      "kernel": {"_type": "choice", "_value": [5, 7, 9, 11]},
      "pool_size": {"_type": "choice", "_value": [3, 5, 7]}
    }
  ]
},
"fe_slot_3": {
  "_type": "choice",
  "_value": [
    {"_name": "none"},
    {
      "_name": "simple",
      "filters": {"_type": "choice", "_value": [8, 16, 32]},
      "kernel": {"_type": "choice", "_value": [5, 7, 9, 11]}
    },
    {
      "_name": "with_pool",
      "filters": {"_type": "choice", "_value": [8, 16, 32]},
      "kernel": {"_type": "choice", "_value": [5, 7, 9, 11]},
      "pool_size": {"_type": "choice", "_value": [3, 5, 7]}
    }
  ]
},
```

Next, we define decision maker hyperparameters:

```
"l1_size": {
    "_type": "choice", "_value": [512, 1024, 2048]},
  "l2_size": {
    "_type": "choice", "_value": [0, 512, 1024]},
  "dropout_rate": {
    "_type": "choice", "_value": [0.3, 0.5, 0.7]},
```

And `learning_rate` hyperparameter finishes the search space:

```
  "learning_rate": {
    "_type": "choice", "_value": [0.001, 0.0001]}
}
```

We have just defined rather nontrivial search space. Let's hope that the result of the experiment will meet our expectations and the resulting architecture will perfectly solve the problem of human and horses classification. The next step is to make TensorFlow and PyTorch implementations of the LeNet Evolution model.

TensorFlow LeNet Evolution Implementation

Listing 2-22 presents the **LeNet Evolution** model's implementation using TensorFlow.

We import necessary modules:

Listing 2-22. LeNet Upgrade. TensorFlow implementation. ch2/lenet_to_alexnet/tf_lenet_evolution.py

```
import tensorflow as tf
from tensorflow.keras import Model
from tensorflow.keras.layers import (
    Conv2D, Dense,
    Dropout, Flatten, MaxPool2D, ReLU,
)
from tensorflow.keras.optimizers import Adam
from ch2.utils.datasets import hoh_dataset
from ch2.utils.tf_utils import TfNniIntermediateResult
```

LeNet Evolution model is initialized with four hyperparameters:

```
class TfLeNetEvolution(Model):

    def __init__(
            self,
            feat_ext_sequences,
            l1_size,
            l2_size,
            dropout_rate
    ):
        super().__init__()
```

Model's layer stack is filled dynamically depending on hyperparameters:

```
        layer_stack = []
```

First, we define feature extraction sequences:

```
        for fe_seq in feat_ext_sequences:
            if fe_seq['type'] in ['simple', 'with_pool']:
                # Constructing Feature Extraction Sequence
                layer_stack.append(
                    Conv2D(
                        filters = fe_seq['filters'],
                        kernel_size = fe_seq['kernel']
                    )
                )
                if fe_seq['type'] == 'with_pool':
                    layer_stack.append(
                        MaxPool2D(
                            pool_size = fe_seq['pool_size']
                        )
                    )
                layer_stack.append(ReLU())

        layer_stack.append(Flatten())
```

Next, we construct decision maker component:

```
layer_stack.append(
    Dense(
        units = l1_size,
        activation = 'relu'
    )
)
layer_stack.append(
    Dropout(rate = dropout_rate)
)
```

Additional linear layer is added if `l2_size` is greater than zero:

```
# Optional Linear Layer
if l2_size > 0:
    layer_stack.append(
        Dense(
            units = l2_size,
            activation = 'relu'
        )
    )
    layer_stack.append(
        Dropout(rate = dropout_rate)
    )
```

Final classification layer:

```
layer_stack.append(
    Dense(
        units = 2,
        activation = 'softmax'
    )
)
```

And here, we set layer sequence to the model:

```
self.seq = tf.keras.Sequential(layer_stack)
```

Model execution method is trivial sequence layer call:

```
def call(self, x, **kwargs):
    y = self.seq(x)
    return y
```

As before, we use the `Adam` optimizer with cross-entropy as the loss function:

```
def train(self, learning_rate, batch_size, epochs):
    self.compile(
        optimizer = Adam(learning_rate = learning_rate),
        loss = 'sparse_categorical_crossentropy',
        metrics = ['accuracy']
    )

    (x_train, y_train), _ = hoh_dataset()

    intermediate_cb = TfNniIntermediateResult('accuracy')
    self.fit(
        x_train,
        y_train,
        batch_size = batch_size,
        epochs = epochs,
        verbose = 0,
        callbacks = [intermediate_cb]
    )
```

Model testing:

```
def test(self):
    (_, _), (x_test, y_test) = hoh_dataset()
    loss, accuracy = self.evaluate(x_test, y_test, verbose = 0)
    return accuracy
```

Since `LeNetEvolutionModel` is done, we can implement the NNI trial script using Listing 2-23.

We import necessary modules and pass code root directory to system path:

Listing 2-23. NNI trial script with TensorFlow LeNetEvolution implementation. ch2/lenet_to_alexnet/tf_trial.py

```
import os
import sys
import nni

# We use relative import for user-defined modules
# For NNI use relative import for user-defined modules
SCRIPT_DIR = os.path.dirname(os.path.abspath(__file__)) + '/../..'
sys.path.append(SCRIPT_DIR)

from ch2.lenet_to_alexnet.tf_lenet_evolution import TfLeNetEvolution
```

The `trial` method initializes the model, trains it, tests it, and returns the NNI metric:

```
def trial(hparams):
```

Feature extraction hyperparameters are converted to universal form:

```
    feat_ext_sequences = []
    for k, v in hparams.items():
        if k.startswith('fe_slot_'):
            v['type'] = v['_name']
            feat_ext_sequences.append(v)
```

Model initialization:

```
    model = TfLeNetEvolution(
        feat_ext_sequences = feat_ext_sequences,
        l1_size = hparams['l1_size'],
        l2_size = hparams['l2_size'],
        dropout_rate = hparams['dropout_rate']
    )
```

Here, we train the model during 50 epochs and fixed batch_size = 16:

```
model.train(
    batch_size = 16,
    learning_rate = 0.001,
    epochs = 50
)
```

Testing the model:

```
accuracy = model.test()
```

And after, we return accuracy metric to NNI tuner:

```
# send final accuracy to NNI
nni.report_final_result(accuracy)
```

Next, we define the main entry point for the trial:

```
if __name__ == '__main__':

    # Manual HyperParameters
    hparams = {
        'fe_slot_1':    {
            '_name':   'simple',
            'filters': 16,
            'kernel':  7
        },
        'fe_slot_2':    {
            '_name':      'with_pool',
            'filters':    8,
            'kernel':     5,
            'pool_size': 5
        },
        'fe_slot_3':    {
            '_name':      'with_pool',
            'filters':    8,
            'kernel':     5,
            'pool_size': 3
        },
```

```
        'l1_size':      1024,
        'l2_size':      512,
        'dropout_rate': .3,
        'learning_rate': 0.001
    }

    # NNI HyperParameters
    # Run safely without NNI Experiment Context
    nni_hparams = nni.get_next_parameter()
    hparams.update(nni_hparams)

    trial(hparams)
```

Remember that a trial script can be executed in stand-alone mode, so you can run a ch2/lenet_to_alexnet/tf_trial.py to test its execution with custom parameters.

PyTorch LeNet Evolution Implementation

Section, we will examine the **LeNet Evolution** model's implementation using PyTorch Lightning. PyTorch Lightning is a seamless PyTorch wrapper that helps eliminate the PyTorch code boilerplate. It is more concise and better suited for such tasks. Listing 2-24 demonstrates the LeNet Evolution model based on PyTorch Lightning.

We import necessary modules:

Listing 2-24. LeNet Evolution. PyTorch Lightning implementation. ch2/lenet_to_alexnet/pt_lenet_evolution.py

```
import nni
import torch
import torch.nn.functional as F
import pytorch_lightning as pl
from sklearn.metrics import accuracy_score
from torch import nn
import numpy as np
from torch.utils.data import DataLoader
from ch2.utils.datasets import hoh_dataset
from ch2.utils.pt_utils import SimpleDataset
```

LeNet Evolution model is initialized with all five hyperparameters. PyTorch Lightning model encapsulates initialization and training logic in the same class, so we pass all hyperparameters at once:

```
class PtLeNetEvolution(pl.LightningModule):

    def __init__(
            self,
            feat_ext_sequences,
            l1_size,
            l2_size,
            dropout_rate,
            learning_rate
    ) -> None:

        super().__init__()
```

`learning_rate` and `dropout_rate` hyperparameters are stored explicitly:

```
        self.lr = learning_rate
        self.dropout_rate = dropout_rate
        self.save_hyperparameters()
```

The first step is to create a layer sequence for the feature extraction component dynamically:

```
        fe_stack = []

        # Input size of next conv layer is out_channels of previous one
        in_dim = 3

        for fe_seq in feat_ext_sequences:
            if fe_seq['type'] in ['simple', 'with_pool']:
                fe_stack.append(
                    nn.Conv2d(
                        in_dim,
                        out_channels = fe_seq['filters'],
                        kernel_size = fe_seq['kernel'],
                        bias = False
                    )
```

```
        )
        if fe_seq['type'] == 'with_pool':
            fe_stack.append(
                nn.MaxPool2d(
                    kernel_size = fe_seq['pool_size']
                )
            )
        fe_stack.append(nn.ReLU())
        in_dim = fe_seq['filters']

self.fe_stack = nn.Sequential(*fe_stack)
```

The next step is to create a layer sequence for the decision maker component:

```
# Lazy fc1 Layer Initialization
self.fc1__in_features = 0
self._fc1 = None
```

Additional linear layer is added if `l2_size` is greater than zero:

```
if l2_size > 0:
    self.fc2 = nn.Sequential(
        nn.Linear(l1_size, l2_size),
        nn.ReLU(),
        nn.Dropout(dropout_rate)
    )
    self.fc3 = nn.Linear(l2_size, 2)
else:
    self.fc2 = nn.Identity()
    self.fc3 = nn.Linear(l1_size, 2)
```

Here, we again utilize the familiar lazy layer initialization pattern:

```
@property
def fc1(self):
    if self._fc1 is None:
        self._fc1 = nn.Sequential(
            nn.Linear(
                self.fc1__in_features,
```

```
                self.hparams['l1_size']
            ),
            nn.ReLU(),
            nn.Dropout(self.dropout_rate)
        )
    return self._fc1
```

Model execution method:

```
def forward(self, x):
    # calling feature extraction layer sequence
    x = self.fe_stack(x)
    # Flatting all dimensions but batch-dimension
    self.fc1__in_features = np.prod(x.shape[1:])
    x = x.view(-1, self.fc1__in_features)
    x = self.fc1(x)
    x = self.fc2(x)
    x = self.fc3(x)
    return F.log_softmax(x, dim = 1)
```

We use the Adam optimizer with the learning_rate hyperparameter:

```
def configure_optimizers(self):
    return torch.optim.Adam(
        self.parameters(),
        lr = self.lr
    )
```

Training and test methods use cross-entropy loss function:

```
def training_step(self, batch, batch_idx):
    x, y = batch
    p = self(x)
    loss = F.cross_entropy(p, y)
    self.log("train_loss", loss, prog_bar = True)
    nni.report_intermediate_result(loss.item())
    return loss
```

```
def test_step(self, batch, batch_idx):
    x, y = batch
    p = self(x)
    loss = F.cross_entropy(p, y)
    self.log('test_loss', loss, prog_bar = True)
    return loss
```

The following method performs the training process on the training dataset and tests the trained model on the test dataset:

```
def train_and_test_model(self, batch_size, epochs):
```

Training and testing datasets are prepared:

```
    (x_train, y_train), (x_test, y_test) = hoh_dataset()
    x_train = torch.from_numpy(x_train).float()
    y_train = torch.from_numpy(y_train).long()
    x_test = torch.from_numpy(x_test).float()
    y_test = torch.from_numpy(y_test).long()

    x_train = torch.permute(x_train, (0, 3, 1, 2))
    x_test = torch.permute(x_test, (0, 3, 1, 2))

    # Dataset to DataLoader
    train_ds = SimpleDataset(x_train, y_train)
    test_ds = SimpleDataset(x_test, y_test)

    train_loader = DataLoader(train_ds, batch_size)
    test_loader = DataLoader(test_ds, batch_size)
```

PyTorch Lightning trainer:

```
    trainer = pl.Trainer(
        max_epochs = epochs,
        checkpoint_callback = False
    )
```

Model training:

```
    trainer.fit(self, train_loader)
```

And finally, we test the trained model:

```
    test_loss = trainer.test(self, test_loader)

    output = self(x_test)
    predict = output.argmax(dim = 1, keepdim = True)
    accuracy = round(accuracy_score(predict, y_test), 4)

    return accuracy
```

Since `LeNetEvolutionModel` is done, we can implement the NNI trial script using Listing 2-15.

We import necessary modules and pass code root directory to system path:

Listing 2-25. NNI trial script with PyTorch Lightning LeNetEvolution implementation. ch2/lenet_to_alexnet/pt_trial.py

```
import os
import sys
import nni

# We use relative import for user-defined modules
# For NNI use relative import for user-defined modules
SCRIPT_DIR = os.path.dirname(os.path.abspath(__file__)) + '/../..'
sys.path.append(SCRIPT_DIR)

from ch2.lenet_to_alexnet.pt_lenet_evolution import PtLeNetEvolution
```

The `trial` method initializes the model, trains it, tests it, and returns the NNI metric:

```
def trial(hparams):
```

Feature extraction hyperparameters are converted universal to form:

```
    feat_ext_sequences = []
    for k, v in hparams.items():
        if k.startswith('fe_slot_'):
            v['type'] = v['_name']
            feat_ext_sequences.append(v)
```

Model initialization:

```
model = PtLeNetEvolution(
    feat_ext_sequences = feat_ext_sequences,
    l1_size = hparams['l1_size'],
    l2_size = hparams['l2_size'],
    dropout_rate = hparams['dropout_rate'],
    learning_rate = hparams['learning_rate']
)
```

Next, we train the model during 50 epochs and fixed `batch_size` = 16 and test it in the same method:

```
accuracy = model.train_and_test_model(
    batch_size = 16,
    epochs = 50
)
```

And after, we return accuracy metric to NNI tuner:

```
# send final accuracy to NNI
nni.report_final_result(accuracy)
```

Next, we define the main entry point for the trial:

```
if __name__ == '__main__':
    # Manual HyperParameters
    hparams = {
        'fe_slot_1':    {
            '_name':   'simple',
            'filters': 16,
            'kernel':  7
        },
        'fe_slot_2':    {
            '_name':      'with_pool',
            'filters':    8,
            'kernel':     5,
            'pool_size': 5
        },
```

```
        'fe_slot_3':    {
            '_name':     'with_pool',
            'filters':   8,
            'kernel':    5,
            'pool_size': 3
        },
        'l1_size':      1024,
        'l2_size':      512,
        'dropout_rate': .3,
        'learning_rate': 0.001
    }

    # NNI HyperParameters
    # Run safely without NNI Experiment Context
    nni_hparams = nni.get_next_parameter()
    hparams.update(nni_hparams)

    trial(hparams)
```

Remember that a trial script can be executed in stand-alone mode, so you can run a ch2/lenet_to_alexnet/pt_trial.py to test its execution with custom parameters.

Performing LeNet Evolution HPO Experiment

And so we come to the climax of this section. We can consider this experiment a full-fledged scientific study that will create a unique deep learning model for a classification problem on a specific dataset. It will take the best principles of pattern recognition from the LeNet model. The LeNet Evolution model attempts to move from the LeNet model to a more complex one. Let's define the experiment configuration and run it finally. Listing 2-26 contains the configuration for the LeNet Evolution HPO Experiment.

Listing 2-26. LeNet Evolution HPO Experiment configuration. ch2/lenet_to_alexnet/config.yml

```
trialConcurrency: 1
```

We will limit the number of trials:

```
maxTrialNumber: 400
searchSpaceFile: search_space.json
trialCodeDirectory: .
```

Uncomment PyTorch trial line to run the experiment using PyTorch implementation:

```
trialCommand: python3 tf_trial.py
#trialCommand: python3 pt_trial.py
```

GridSearch Tuner cannot be used for search spaces that utilize nested choice, so we pick Random Search Tuner:

```
tuner:
  name: Random
trainingService:
  platform: local
```

The experiment can be run as follows:

```
nnictl create --config ch2/lenet_to_alexnet/config.yml
```

Note Duration ~ 18 hours on Intel Core i7 with CUDA (GeForce GTX 1050)

Best trial hyperparameters returned by the experiment are listed in Table 2-6.

Table 2-6. *LeNet Evolution best hyperparameters*

Name	Values
`fe_slot_1`	• `with_pool`: • `filters`: 32 • `kernel`: 7 • `pool_size`: 5
`fe_slot_2`	• `with_pool`: • `filters`: 8 • `kernel`: 11 • `pool_size`: 5
`fe_slot_3`	• `simple`: • `filters`: 8 • `kernel`: 7
`l1_size`	1024
`l2_size`	512
`dropout_rate`	0.3
`learning_rate`	0.0001

The best trial demonstrated a **0.9941** accuracy on test dataset, which is an excellent result. We have indeed managed to build a model that distinguishes complex colored objects with a very high degree of accuracy. This is a good development! The reader may wonder: *Why is this section called From LeNet to AlexNet*? Well, it's time to answer it. AlexNet is the name of a convolutional neural network architecture that won the 2012 Image Recognition competition. AlexNet classified images into 1000 different classes. At that time, it was a pretty advanced deep learning model. Let's now compare three models: the LeNet model, the model we constructed in this section using HPO techniques (LeNet Evolution), and the AlexNet model.

Figure 2-22 shows that the model we built in this section for humans and horses classification is somewhere between the original LeNet model and the AlexNet model. Our model shows a remarkable test result of **99.41%** accuracy. But most importantly, it was built entirely automatically with the help of HPO techniques and the NNI tool! We did not do any complex calculations or analytical analysis. We have just constructed a flexible LeNet Evolution model whose architecture depended on the passed hyperparameters. And as a result, we got a unique model that is fully adapted to solving a specific task. These results confirm the promise of HPO's approach to solving deep learning problems.

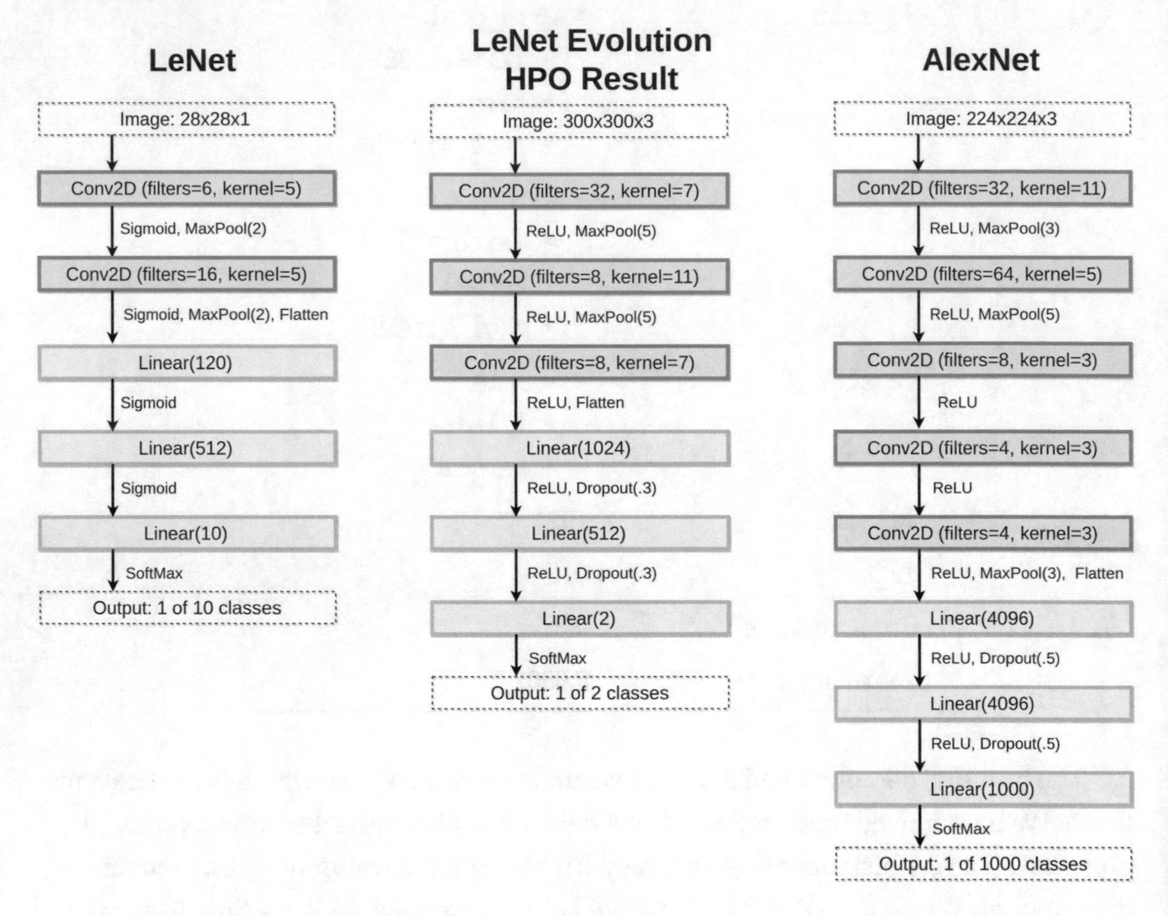

Figure 2-22. *LeNet, LeNet Evolution, AlexNet*

Summary

In this chapter, we started the HPO study. We studied how to create NNI experiments and solve practical problems. We managed to optimize the original LeNet model for handwritten digit recognition, upgrade the LeNet model using ReLU and dropout techniques, and construct a new complex color pattern recognition model based on the existing LeNet model. The results that we have obtained demonstrate the promise of AutoDL. In the next chapter, we will continue to study HPO and dive into the more advanced NNI usage in Hyperparameter Optimization problems.

CHAPTER 3

Hyperparameter Optimization Under Shell

In the previous chapter, we saw that simple HPO techniques could produce very impressive results. Hyperparameter Optimization not only optimizes a specific model for a dataset but can even construct new architectures. But the fact is that we have used an elementary set of tools for HPO tasks so far. Indeed, up to this point, we have only used the primitive Random Search Tuner and Grid Search Tuner. We learned from the previous chapter that search spaces could contain millions and hundreds of millions of parameters. And if we had unlimited time, we could always use the Grid Search Tuner. But unfortunately, this approach is not applicable in reality. We need Tuners that strike a good balance between speed and quality in finding the best hyperparameters. Another helpful technique is Early Stopping algorithms. Early Stopping algorithms analyze the model training process based on intermediate results and decide whether to continue training or stop it to save time.

This chapter will study various HPO Tuners and tell about their basic features. We will explore the use of Early Stopping algorithms that speed up the experiment stopping trials with an unpromising training process. And also, consider creating a custom HPO Tuner for a particular task. This chapter will greatly enhance the practical application of the Hyperparameter Optimization approach.

Tuners

We begin this chapter by examining the various HPO Tuners. As you remember, Tuner receives metrics from Trial after evaluating a particular search space parameter. Based on the existing result history of all completed Trials, Tuner decides which hyperparameter configuration to test next. The main task of the Tuner is to find the best

I. Gridin, *Automated Deep Learning Using Neural Network Intelligence*,
https://doi.org/10.1007/978-1-4842-8149-9_3

hyperparameters as quickly as possible. The choice of a suitable tuner can significantly improve the result of an experiment. Let's take a closer look at how Random Search Tuner and Grid Search Tuner act.

Consider a two-variable black-box function which is expressed in Listing 3-1.

Listing 3-1. Black-box function. ch3/bbf/f1.py

```
from ch3.bbf.utils import discrete, noise, scatter_plot
def black_box_f1(x, y):
    z = - 10 * (pow(x, 5) / (3 * pow(x * x * x / 4 + 1, 2) + pow(y, 4) +
    10) + pow(x * y / 2, 2) / 1000)
    d = discrete(z, .8)
    r = d + noise(x, y, scale = 8)
    d = discrete(r, .2)
    return r
```

I deliberately do not give an analytical formula for this function. `black_box_f1` is just a black-box function, and we know nothing about its internal logic. In real life, black-box functions have the following properties:

- They are not continuous
- They have random noise

All black-box functions that we will examine in this chapter will satisfy these properties. But anyway, we can cheat a little and plot this function:

```
if __name__ == '__main__':
    scatter_plot(black_box_f1, [-10, 10], [-10, 10])
```

Figure 3-1 shows that the red area is where the black-box function `black_box_f1` reaches its maximum values.

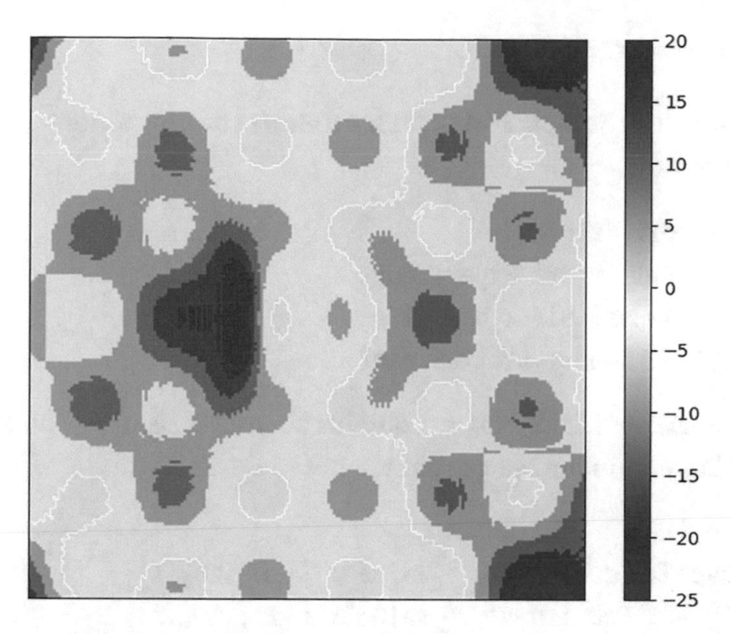

***Figure 3-1.** Black-box function plot*

Note In this chapter, we will examine only problems of finding the maxima of the black-box function ***f***. The problem of finding the maxima of the function f is equivalent to the problem of finding the minima of the function -***f***.

Of course, it would be great to be able to visualize a black-box function before research, but there are two main problems here:

- Most of the black-box functions have a large number of variables.
- Calculating one function value with specific parameters can take minutes and even hours.

Therefore, of course, the choice of a Tuner has great importance. Let's take a look at how Random Search Tuner explores the `black_box_f1` function. We are implementing an embedded experiment to visualize the trial parameters that the tuner has selected during the experiment in Listing 3-2.

We import necessary modules:

Listing 3-2. Random Search Tuner. ch3/tuners/random_tuner/run_experiment.py

```
from pathlib import Path
from nni.experiment import Experiment
from ch3.bbf.f1 import black_box_f1
from ch3.bbf.utils import scatter_plot
```

The search space for `black_box_f1` contains all integer pairs in [-10, 10] × [-10, 10]. There are 441 elements in the search space.

```
search_space = {
    "x": {"_type": "quniform", "_value": [-10, 10, 1]},
    "y": {"_type": "quniform", "_value": [-10, 10, 1]}
}
```

The experiment will have 100 trials:

```
experiment = Experiment('local')
experiment.config.experiment_name = 'Random Tuner'
experiment.config.trial_concurrency = 4
experiment.config.max_trial_number = 100
experiment.config.search_space = search_space
experiment.config.trial_command = 'python3 trial.py'
experiment.config.trial_code_directory = Path(__file__).parent
```

We pick Random Search Tuner

```
experiment.config.tuner.name = 'Random'
```

and start the experiment:

```
http_port = 8080
experiment.start(http_port)
```

Next, we define the main event loop:

```
while True:
    if experiment.get_status() == 'DONE':
```

When the experiment is finished, we display all the trials that were created during the experiment:

```
search_data = experiment.export_data()
trial_params = [trial.parameter for trial in search_data]

# Visualizing Trial Parameters
scatter_plot(
    black_box_f1, [-10, 10], [-10, 10],
    trial_params, title = 'Random Search'
)

search_metrics = experiment.get_job_metrics()
input("Experiment is finished. Press any key to exit...")
break
```

Let's examine all the trials that Random Search Tuner generated during the experiment in Listing 3-2.

Figure 3-2 shows that the trials generated by Random Search Tuner are simple random dots scattering. In some cases, the dot (trial) may successfully fall into the area of maximum values, but in many cases, the area of maximum values remains unexplored properly.

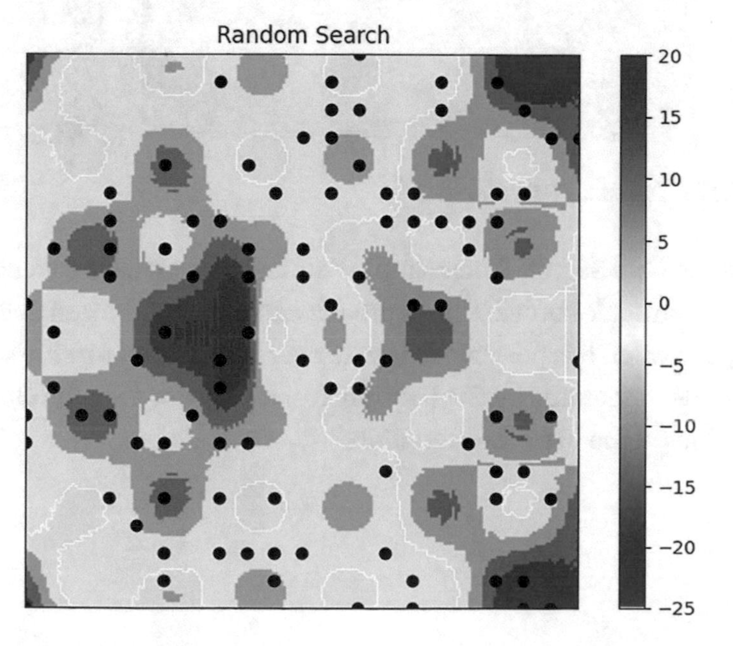

***Figure 3-2.** Random Search Tuner Trials*

Let's now look at the trials that the Grid Search Tuner generates by exploring the `black_box_f1` function in Listing 3-3.

Listing 3-3. Grid Search Tuner. ch3/tuners/grid_tuner/run_experiment.py

```
experiment.config.tuner.name = 'GridSearch'
```

The Grid Search experiment looks much like the Random Search experiment in Listing 3-2. We only use Grid Search Tuner here:

Figure 3-3 demonstrates the trials generated by the Grid Search Tuner.

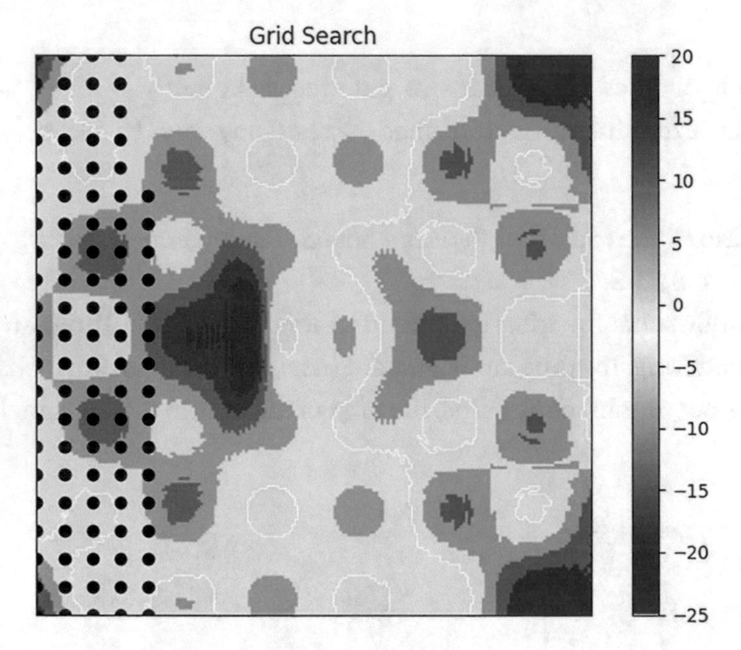

Figure 3-3. *Grid Search Tuner Trials*

We see that the Grid Search Tuner simply iterates through all the values in the search space in a particular order. This approach can be helpful when dealing with a small search space when it is possible to iterate over all the values in the search space. Otherwise, the trials generated by Grid Search Tuner may not even get close to the area of maximum values of the black-box function.

The main problem with Random and Grid tuners is that they don't interact with their trial results in any way. They do not have any "memory" that would allow them to highlight promising areas in the search space and concentrate their search on them. We will now begin to study tuners that have "memory" and which can explore the search space more efficiently.

Evolution Tuner

Evolution Tuner search is based on the principles of natural evolution. It implements two fundamental principles of evolution: selection and mutation. Evolution Tuner initializes a population of a specific size. Each population individual represents a particular set of parameters in the search space. Each individual has a fitness property that indicates the Trial result. We say that individual `A` is better than individual `B`:

- If `A.fitness` > `B.fitness` when the Tuner runs in maximization mode
- If `A.fitness` < `B.fitness` when the Tuner runs in minimization mode

Evolution Tuner takes the best individual from a random pair of individuals and mutates it randomly by replacing the value of its parameter with another value from the search space. After that, the mutated individual replaces the original one, and the process is repeated again. Figure 3-4 illustrates this search principle.

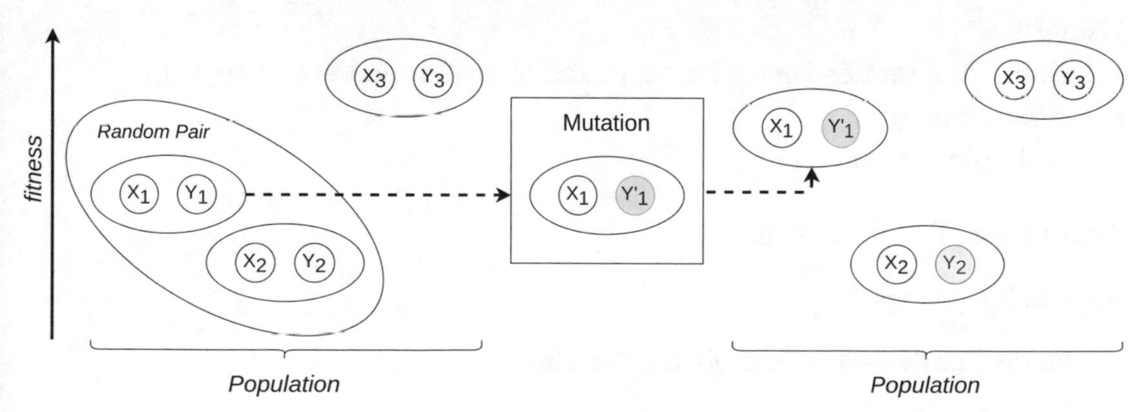

Figure 3-4. *Evolution Tuner*

One of the big problems with Evolution Tuner is that mutation doesn't always improve an individual's fitness. Mutation operation only executes a random change in parameter values, and usually, mutation degrades an individual's performance.

Note The experienced reader may notice that evolutionary algorithms based on the principles of natural selection have another key method – *crossover*. But many studies show that most evolutionary problems can be solved without the *crossover* operation. This implementation of Evolution Tuner does not contain *crossover* operation.

Here is an example of Evolution Tuner configuration:

```
# config.yml
tuner:
  name: Evolution
  classArgs:
    optimize_mode: maximize
    population_size: 100
```

Evolution Tuner supports all search space types: `choice`, `choice(nested)`, `randint`, `uniform`, `quniform`, `loguniform`, `qloguniform`, `normal`, `qnormal`, `lognormal`, and `qlognormal`.

Let's take a look at Evolution Tuner in action optimizing `black_box_f1` function in Listing 3-4.

(Full code is provided in the corresponding file: *ch3/tuners/evolution_tuner/run_experiment.py*.)

Setting population size:

Listing 3-4. Evolution Tuner

```
population_size = 8
```

Picking the Evolution Tuner for the Experiment:

```
experiment.config.tuner.name = 'Evolution'
experiment.config.tuner.class_args['optimize_mode'] = 'maximize'
experiment.config.tuner.class_args['population_size'] = population_size
```

Unlike Grid Search Tuner and Random Search Tuner, an Evolution Tuner has "memory." That is why it is more attractive to analyze the search process progress. We will show the history of the allocation of trial parameters in the search space by generations:

- **1st generation**: From 1st trial to 25th trial
- **2nd generation**: From 26th trial to 50th trial
- **3rd generation**: From 51st trial to 75th trial
- **4th generation**: From 76th trial to 100th trial

This approach will allow us to observe the search progress in action:

```
# Event Loop
while True:
    if experiment.get_status() == 'DONE':
        search_data = experiment.export_data()
```

The history of trials:

```
        trial_params = [trial.parameter for trial in search_data]
```

Splitting trial history by generations:

```
        trial_params_chunks = [
            trial_params[i:i + 25]
            for i in range(0, len(trial_params), 25)
        ]
```

Visualizing each generation:

```
        for i, population in dict(enumerate(trial_params_chunks)).items():
            scatter_plot(
                black_box_f1, [-10, 10], [-10, 10],
                population, title = f'Evolution Generation: {i+1}'
            )
```

Let's analyze how the position of individuals changed during the evolutionary search.

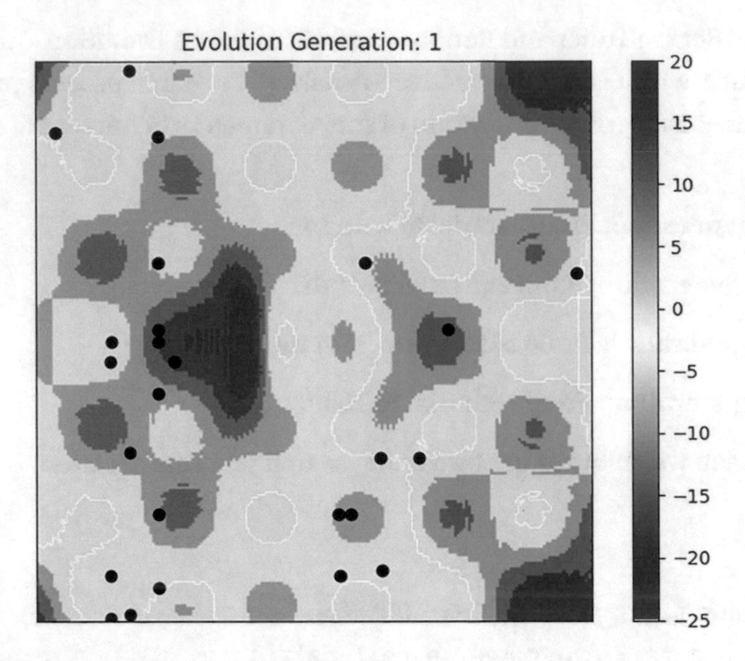

Figure 3-5. *Evolution Tuner. Generation: 1*

Figure 3-5 shows that the allocation of trial parameters is close to random distribution.

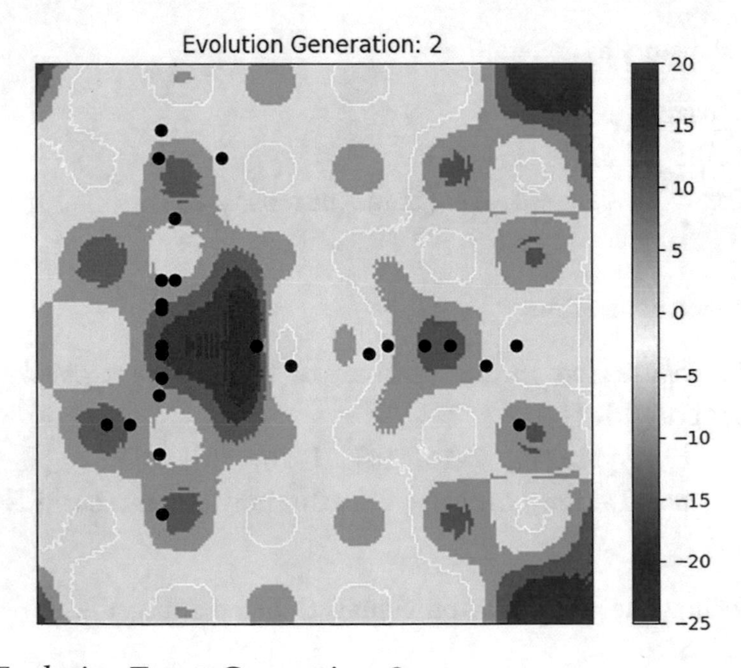

Figure 3-6. *Evolution Tuner. Generation: 2*

In Figure 3-6, we see that most of the individuals are already in the red zone, which means that the population is moving smoothly toward the highest values of the function. And the last generation shown in Figure 3-7 has at least one individual at the top of the red zone.

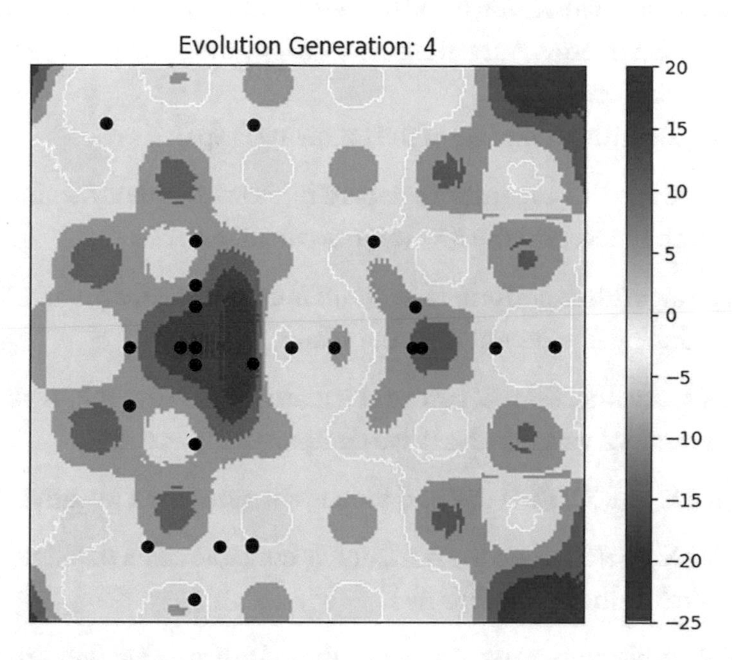

***Figure 3-7.** Evolution Tuner. Generation: 4*

Note Not all built-in NNI tuners support random seed setting. Hence, the experiments are not reproducible. Therefore, the results you get on your local machine may differ from those shown in this chapter. However, the general behavior of tuners remains the same on all machines, so the results of the same tuner on the same search space are similar.

We can consider the Evolution Tuner as a directed random search. It is slightly better than random search but still has many problems due to the too random nature of this algorithm. Evolution Tuner usually requires many trials but is usually selected due to its simplicity.

Anneal Tuner

Anneal Tuner is based on the Simulated Annealing algorithm. Simulated Annealing is a method for solving optimization problems. The algorithm models the physical process of heating a material and slowly lowering the temperature to decrease defects, thus minimizing the system energy. Annealing Tuner uses randomness as part of the search process like Evolution Tuner.

The annealing algorithm consists of the following steps:

1. The annealing algorithm selects a random element X in the search space, and the f(X) value is calculated.

2. The algorithm performs a random mutation on element X producing X' element from the search space.

If X is a real value, then X' can be calculated as $X' = X + \Delta X$, where ΔX is a random variable. Next, we compare f(X') and f(X).

3a. If $f(X') < f(X)$, then the mutation is considered as negative.

3b. If $f(X') \geq f(X)$, then the mutation is considered as positive and X value is updated by X'.

4. If the mutation is negative, then the algorithm calculates the following values:

- *r*: Uniform random value on (0, 1).
- Δ: f(X) - f(X').
- σ is standard deviation of all explored values during the search: $f(X_1), \ldots, f(X_n)$ multiplied by degradation ratio c^i, where *c* is the positive value lower than 1 and *i* is the number of iteration, $\sigma = c^i\, std([f(X_1), \ldots, f(X_n)])$.

Next, we compare *r* and $e^{\frac{\Delta}{\sigma}}$, where *e* is the exponential.

5a. If $r < e^{\frac{\Delta}{\sigma}}$, then the algorithm degrades from X to X': $X \leftarrow X'$.

This is done hoping that it will be possible to reach a new peak in the next iteration and the transition to X' is only an intermediate step. We can consider this as an exploration step that explores an area near X in the search space.

5b. If $r > e^{\frac{\Delta}{\sigma}}$, then the algorithm doesn't update X.

The closer f(X) is to f(X'), the more likely an exploration step would be taken.

Steps from 2 to 5 are repeated *n* times. Figure 3-8 demonstrates the annealing algorithm flow.

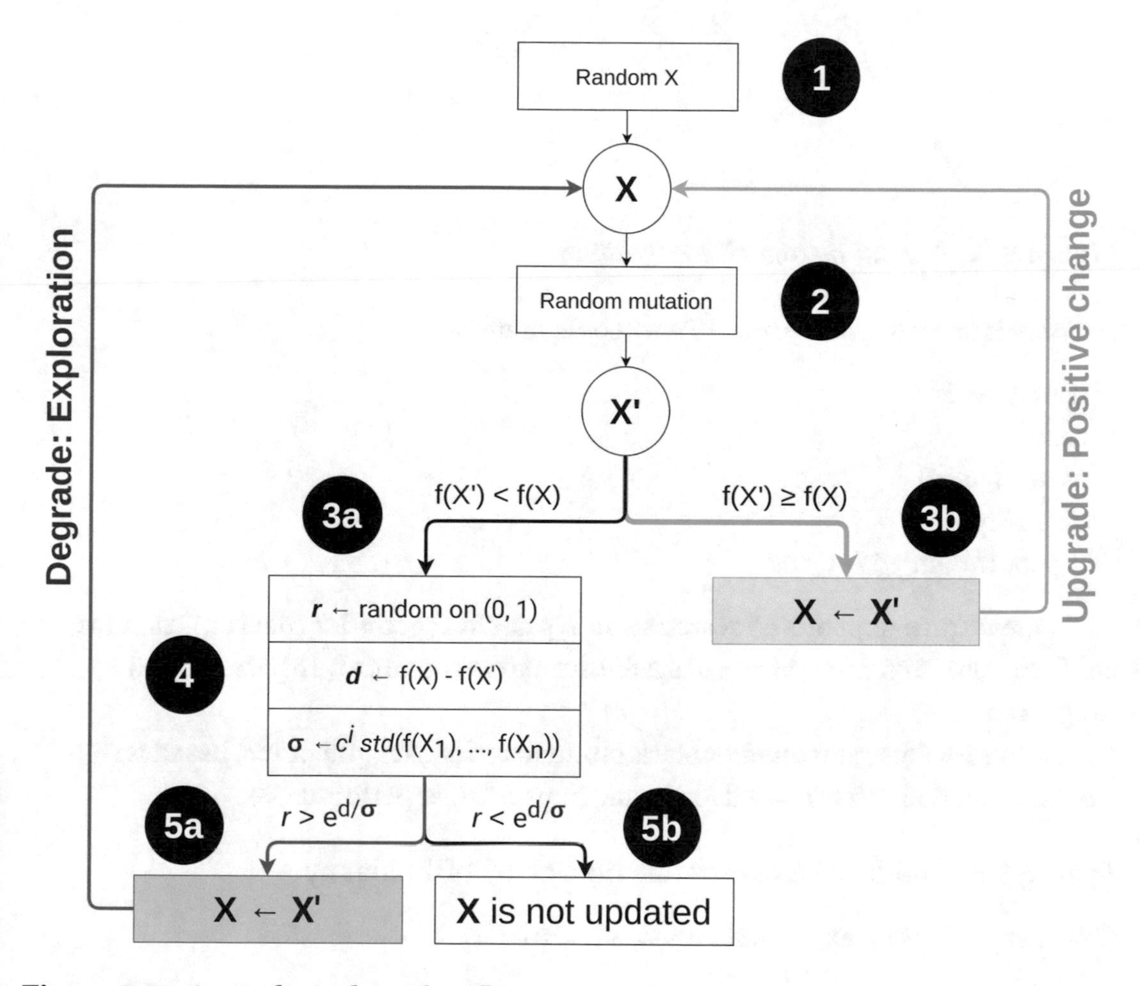

***Figure 3-8.** Annealing algorithm flow*

The essence of the annealing algorithm is to get to the "hills" of the surface of the black-box function *f* and study the area of this "hill." In some cases, the algorithm may descend from "hills" hoping to climb to a higher one. The disadvantage of this algorithm is that it cannot cover large distances between different "hills" of the surface of the function f. Figure 3-9 demonstrates the annealing algorithm in action.

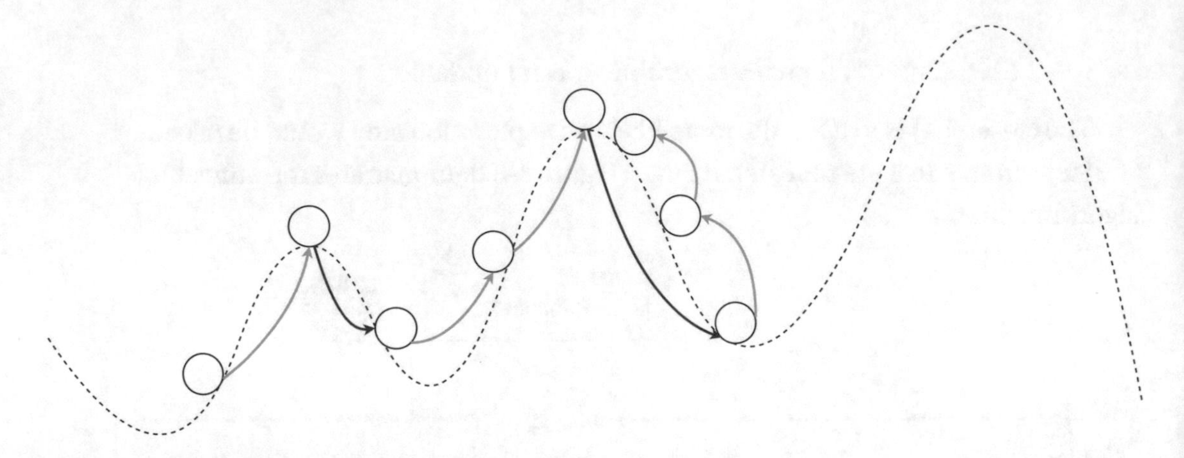

Figure 3-9. *Annealing algorithm in action*

Here is an example of Anneal Tuner configuration:

```
# config.yml
tuner:
  name: Anneal
  classArgs:
    population_size: 100
```

Anneal Tuner supports all search space types: `choice`, `choice(nested)`, `randint`, `uniform`, `quniform`, `loguniform`, `qloguniform`, `normal`, `qnormal`, `lognormal`, and `qlognormal`.

Listing 3-5 illustrates another black-box function `holder_function`, based on Holder's function. We will use it for testing Anneal Tuner performance.

Listing 3-5. Holder's black-box function. ch3/bbf/holder.py

```
from numpy import exp, sqrt, cos, sin, pi

def holder_function(x, y):
    """
    Holder's function
    """
    z = abs(sin(x) * cos(y) * exp(abs(1 - (sqrt(x**2 + y**2) / pi))))
```

```
    d = discrete(z, .8)
    r = d + noise(x, y, scale = 8)
    d = discrete(r, .2)
    return r
```

holder_function can be plotted the following way:

```
from ch3.bbf.utils import scatter_plot, discrete, noise

if __name__ == '__main__':
    scatter_plot(holder_function, [-10, 10], [-10, 10])
```

Figure 3-10 shows the surface of holder_function function. This surface is much more challenging to explore. It has many hills that are evenly spaced on the surface. The highest peaks are in the left corners.

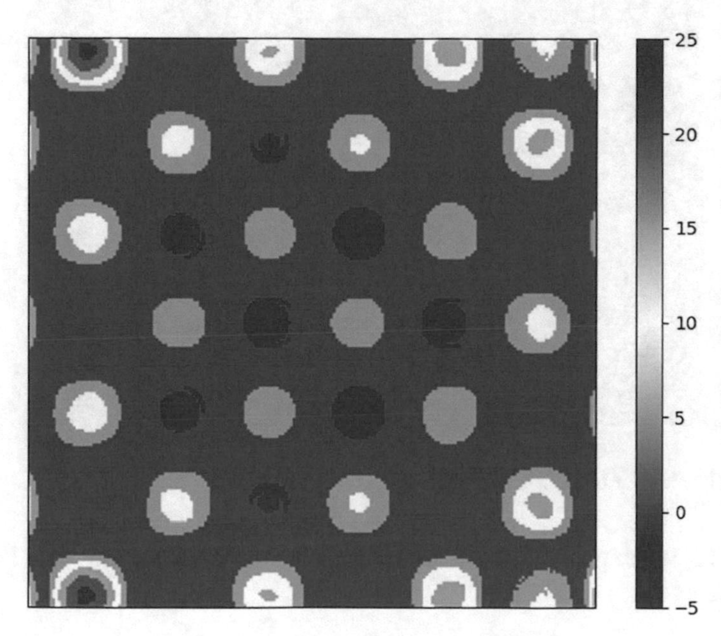

Figure 3-10. *Holder's function*

Let's examinc how Anneal Tuner performs optimizing holder_function in Listing 3-6.

(Full code is provided in the corresponding file: *ch3/tuners/anneal_tuner/run_experiment.py*.)

Picking the Anneal Tuner for the Experiment:

Listing 3-6. Anneal Tuner

```
experiment.config.tuner.name = 'Anneal'
experiment.config.tuner.class_args['optimize_mode'] = 'maximize'
```

After the experiment is completed, we can analyze the progress of the search for Anneal Tuner.

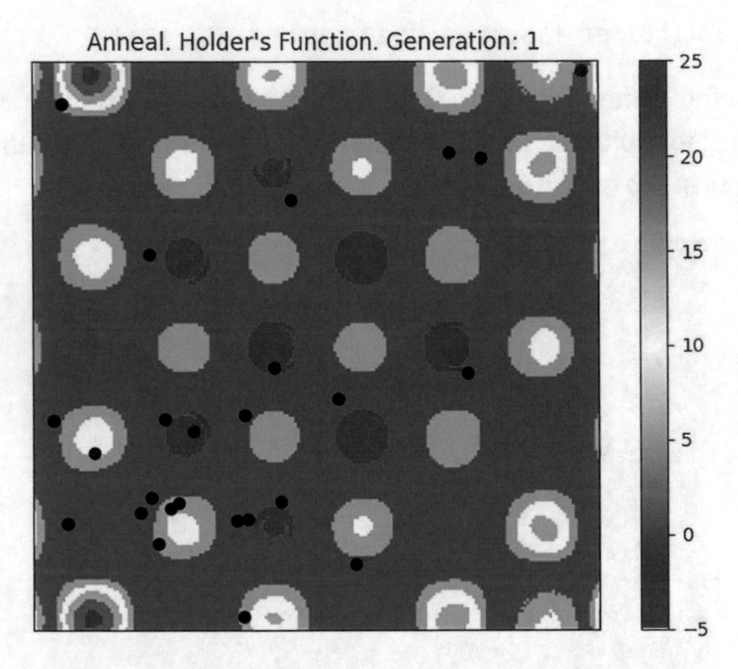

Figure 3-11. *Anneal Tuner. Generation: 1*

In Figure 3-11, we can see that Anneal Tuner is starting to study the "hills" in the lower-left corner.

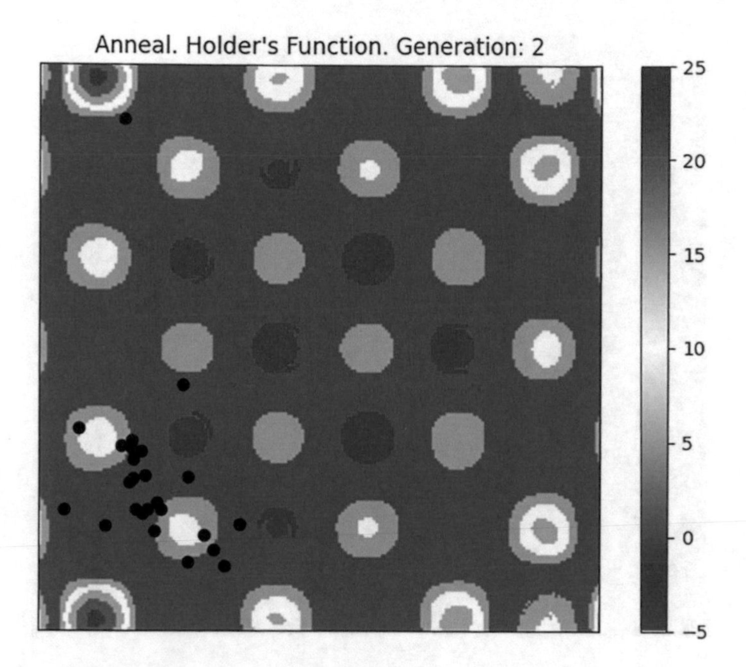

Figure 3-12. *Anneal Tuner. Generation: 2*

Figure 3-12 demonstrates that the second generation of trials is completely focused on the two "hills" in the lower-left corner.

And as we can see in Figure 3-13, Anneal Tuner is completely concentrated on exploring only one "hill." Anneal Tuner found a local maxima but could not find global maxima at the bottom of the left corner, close to the solution Anneal Tuner found.

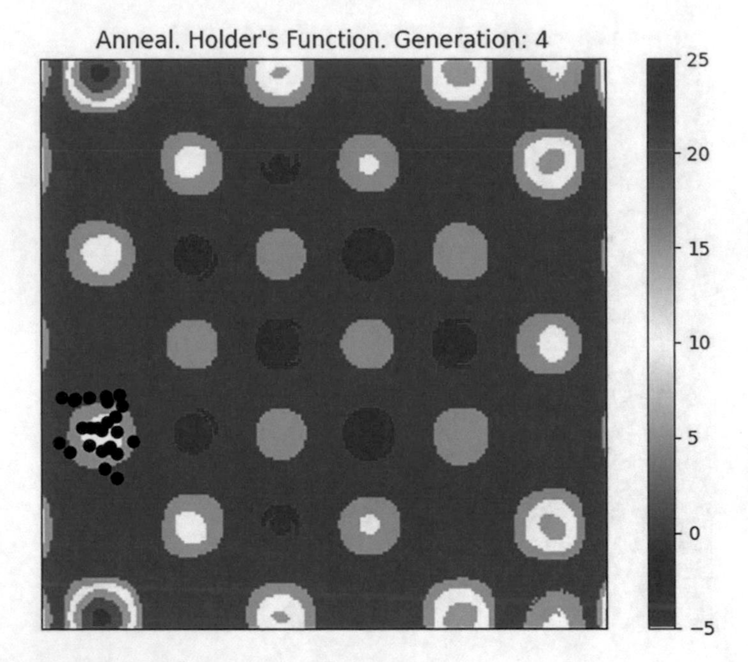

Figure 3-13. *Anneal Tuner. Generation: 4*

Anneal Tuner and Evolution Tuner are variants of directed random search. They are intuitive and straightforward, but they may not always explore the search space effectively. Let's study more advanced tuners based on the Bayesian optimization approach.

Sequential Model-Based Optimization Tuners

In this section, we will examine tuners that are based on Sequential Model-Based Optimization (SMBO). SMBO is a formulation of the Bayesian Optimization approach. SMBO implements the following technique: building a probability model p(y|x) of the black-box function *f* and use it to pick the most promising elements in the search space to evaluate in the black-box function *f*.

Examine the SMBO method in action. Say we have some trial results we obtained: $(x_1, f(x_1))$, $(x_2, f(x_2))$, $(x_3, f(x_3))$, as is shown in Figure 3-14.

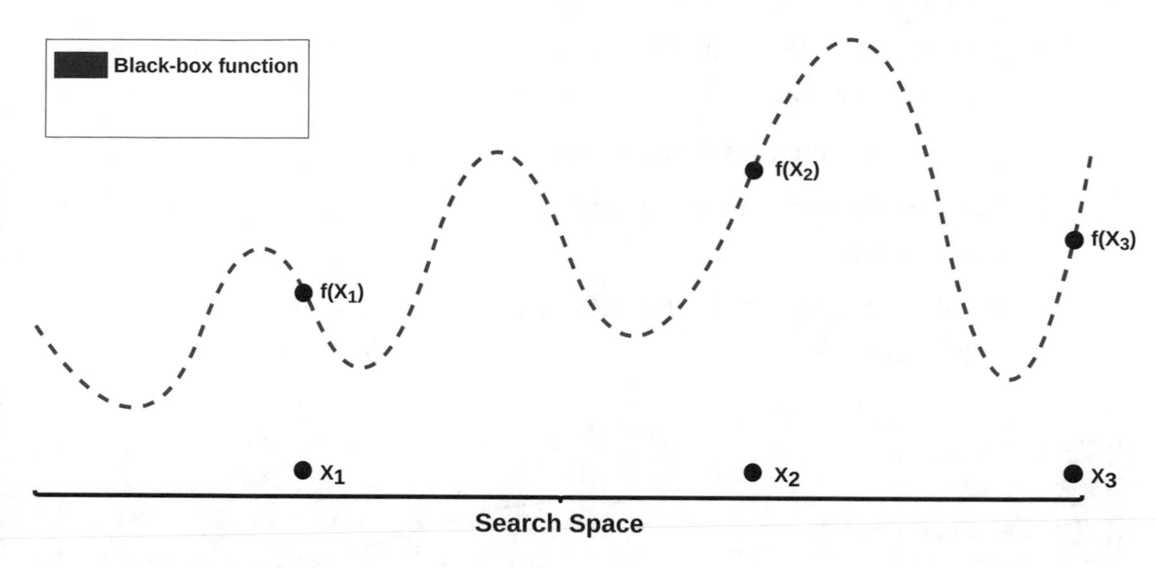

Figure 3-14. *Trial results*

The next step is to create a probability function p(y|x) based on the data: (x_1, f(x_1)), (x_2, f(x_2), (x_3, f(x_3)). p(y|x) is called a "surrogate" for the objective (or black-box) function. The surrogate function determines the probability distribution of the objective (or black-box) function for any element x_i in the search space. This means that for any x_i, we can say that with a *p* probability, the value f(x_i) = y_i lies in the (a, b) interval. This concept is demonstrated in Figure 3-15.

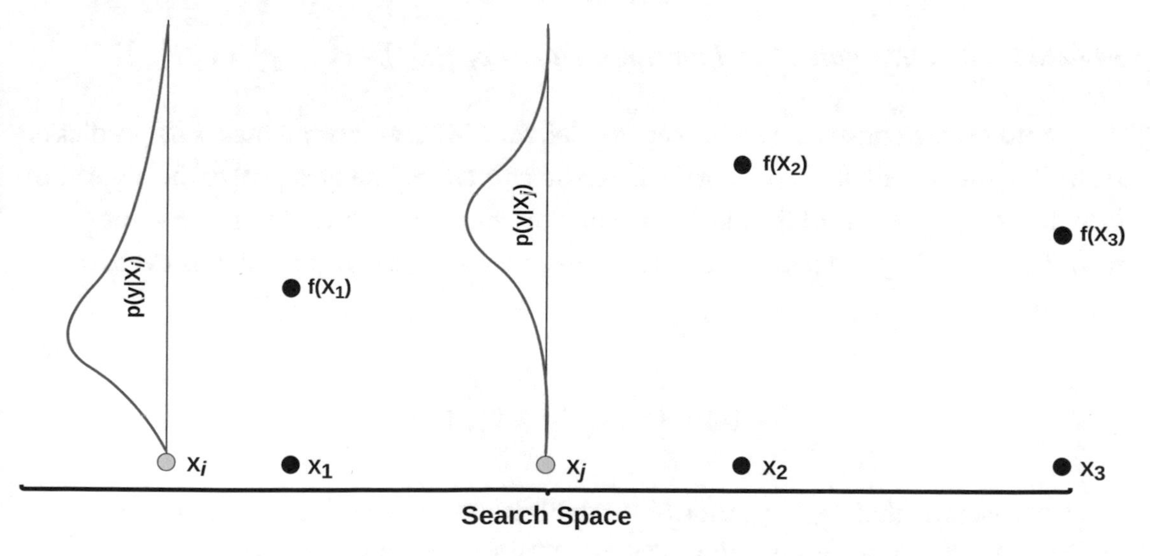

Figure 3-15. *Probability distributions for x_i and x_j*

Having surrogate function p(x|y), we can extrapolate it on the whole search space. Figure 3-16 gives a visual description of the surrogate model:

- Red dashed line shows the actual black-box function.
- Black solid line shows the expected mean of the surrogate function p(y|x).
- Purple dashed line shows the variance of the surrogate function p(y|x).

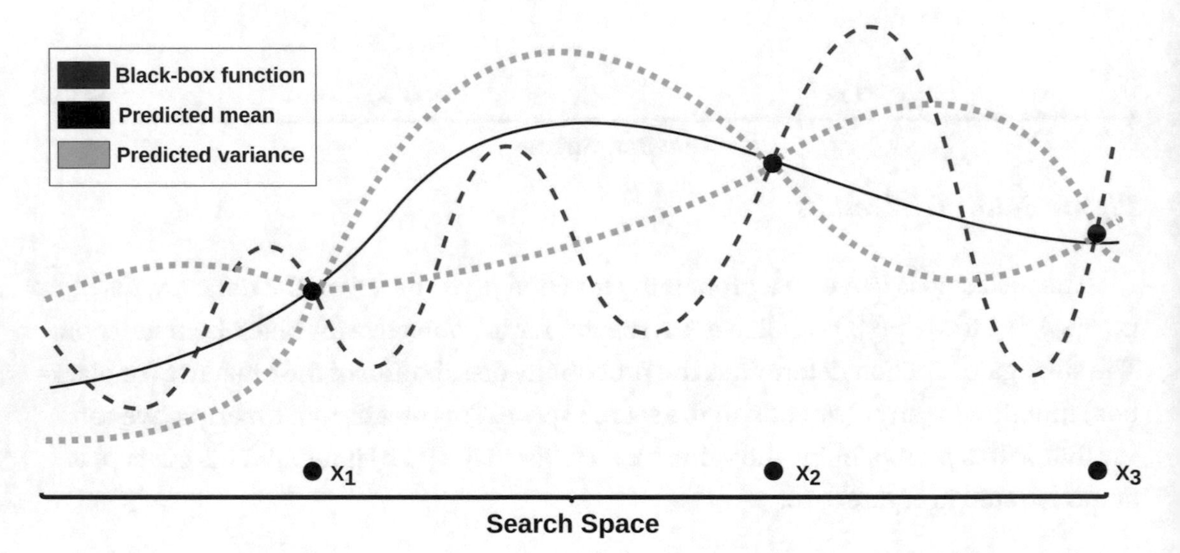

Figure 3-16. *Surrogate model for three trials: $(x_1, f(x_1))$, $(x_2, f(x_2))$, $(x_3, f(x_3))$*

Based on the constructed surrogate model, the SMBO algorithm makes its prediction regarding the potential maxima of the black-box function. The next goal of the algorithm is to find a higher value of the black-box function than the current maximum value $f(x_2)$. The following trial parameters are determined using the Expected Improvement function:

$$EI_{y^*}(x) = \int_{-\infty}^{+\infty} \max\left(y^* - y, 0\right) p\left(y|x\right) dy$$

If we assume that $f(x_2) = y_2$, then SMBO will choose x_4 as the next trial parameter if the $EI_{y2}(x)$ will be maximum with x_4. Figure 3-17 illustrates the next trial selection by SMBO algorithm.

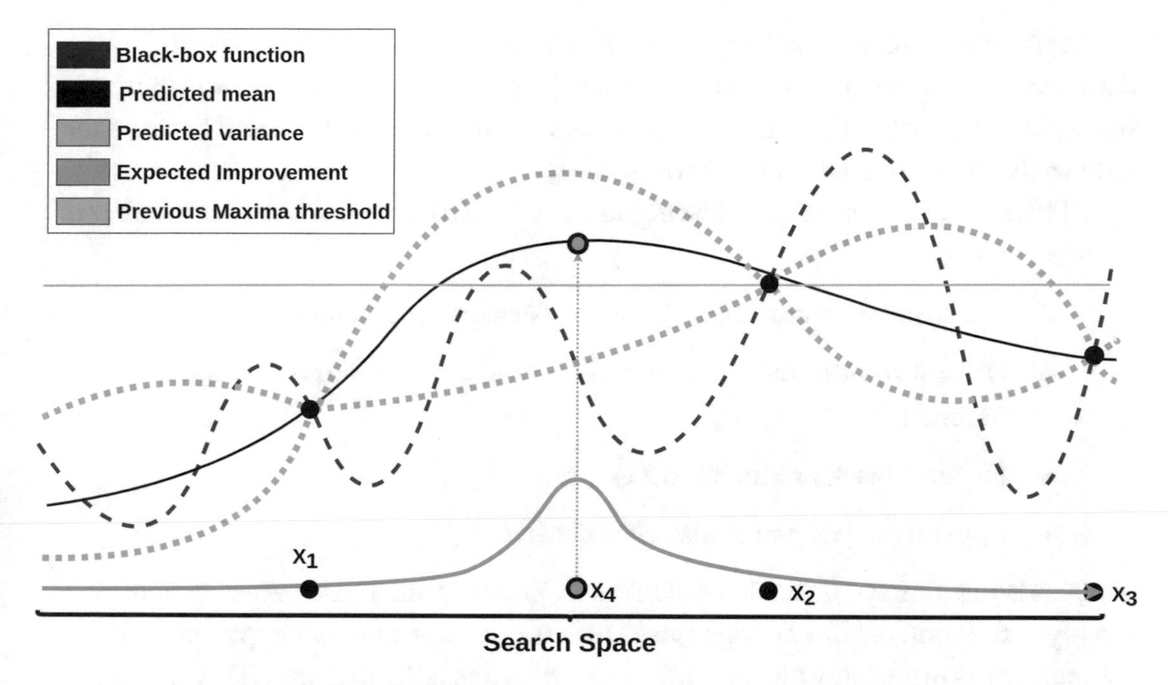

Figure 3-17. *Expected Improvement*

After selecting the x_4 as the next trial value, we evaluate $f(x_4)$ and rebuild the surrogate model concerning the new data: $(x_1, f(x_1))$, $(x_2, f(x_2))$, $(x_3, f(x_3))$, $(x_4, f(x_4))$ as shown in Figure 3-18.

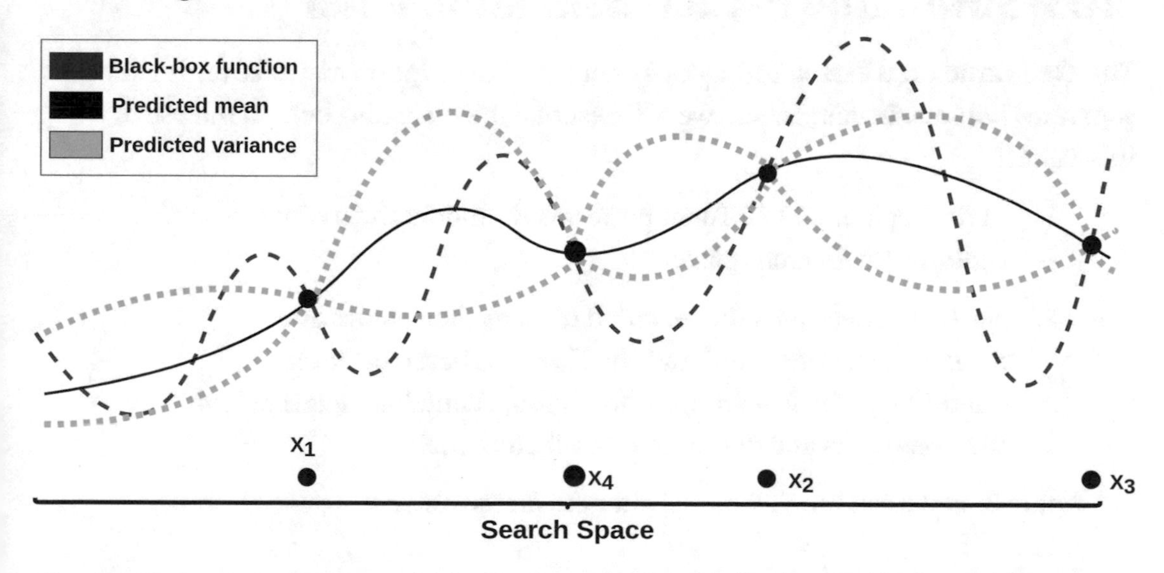

Figure 3-18. *Surrogate model for four trials:* $(x_1, f(x_1))$, $(x_2, f(x_2))$, $(x_3, f(x_3))$, $(x_4, f(x_4))$

SMBO aims to converge the surrogate model to the objective function with more data, which these approaches do by continually updating the surrogate probability model after each objective function evaluation. SMBO Tuners are efficient because they choose the next parameters in an informed way.

SMBO Tuner performs the following steps in a cycle until the maximum number of trials is reached:

- Construct surrogate model based on surrogate probability p(y|x).
- Determine next trial parameter x using the Estimated Improvement function.
- Evaluate black-box function f(x).
- Append *(x, f(x))* pair to historical dataset.

This is the framework for all SMBO Tuners. The only difference between them is the p(y|x) function definition. Different SMBO Tuners have different approaches to estimating the probability function p(y|x) based on a historical dataset. This chapter will cover the following SMBO Tuners: Tree-Structured Parzen Estimator Tuner and Gaussian Process Tuner.

Tree-Structured Parzen Estimator Tuner

The Tree-Structured Parzen Estimator Tuner (TPE) description may take one or several separate chapters. In this section, we will describe the main idea behind the use of this Tuner:

1. In the beginning, TPE Tuner performs *N* random trials while exploring the search space.
2. Next, the Tuner sorts the executed trials by their values and divides them into "good" and "bad" groups based on some quantile - γ. The first group, "good" group, contains trials that gave the best results and the "bad" one, all other trials.

Figure 3-19 depicts the TPE model after the first two steps.

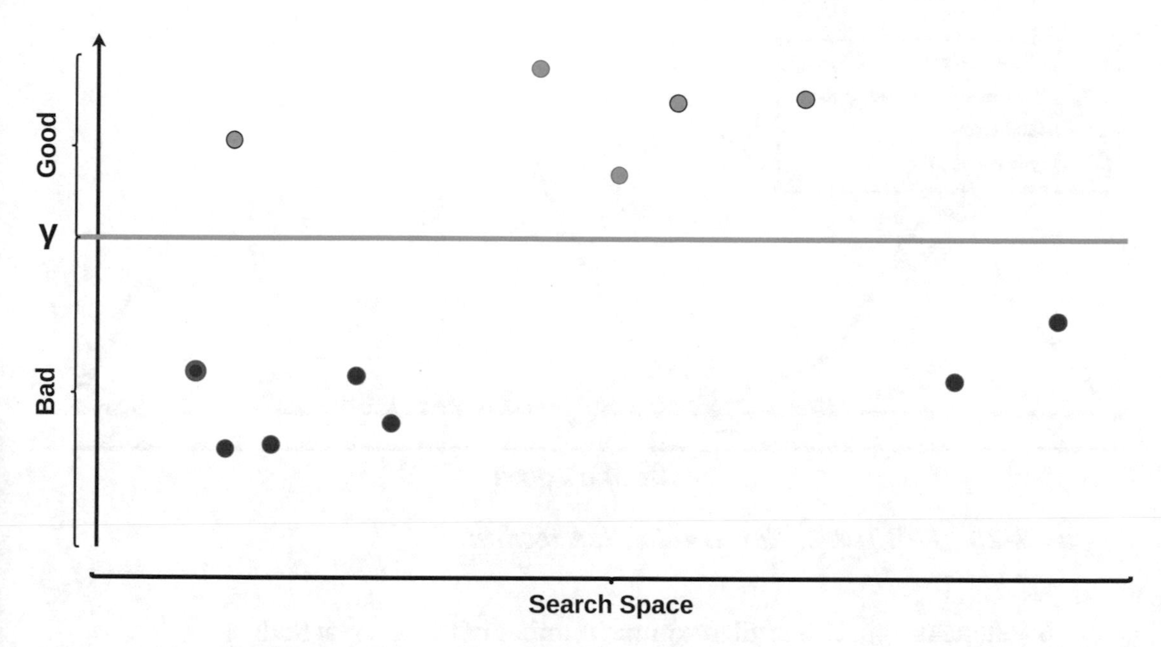

Figure 3-19. *TPE Tuner. "Good" and "bad" separation*

3. Densities $l(x)$ and $g(x)$ are calculated using Parzen Estimators (also known as kernel density estimators) for "bad" and "good" groups, respectively. Parzen Estimators are a simple average of kernels centered on existing data points.

4. After, the TPE tuner generates n random candidates according to the $g(x)$ density function. Each of these candidates is sorted by $g(x)/l(x)$ ratio, and the first one is picked as the next trial. This means that the TPE tuner allows the selection of random candidates in an area where "good" trials are more common. At the same time, all candidates are sorted according to $g(x)/l(x)$, which means that a candidate with a high density of good trials and a low density of bad trials will be selected. This approach strikes a good balance between exploration and exploitation.

Figure 3-20 illustrates the algorithm of selecting the next trial parameter.

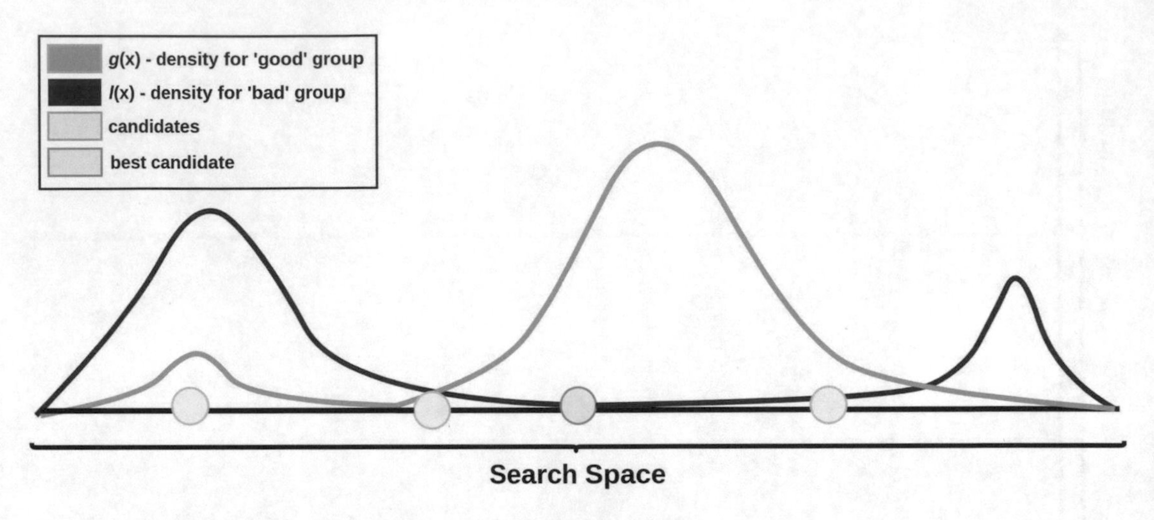

Figure 3-20. *TPE Tuner. Next candidate selection*

5. Repeat steps 2–5 until maximum number of trials is reached.

Here is an example of TPE Tuner configuration:

```
# config.yml
tuner:
  name: TPE
  classArgs:
    optimize_mode: maximize
    seed: 12345
    tpe_args:
      constant_liar_type: 'mean'
      n_startup_jobs: 10
      n_ei_candidates: 20
      linear_forgetting: 100
      prior_weight: 0
      gamma: 0.5
```

The following is the description of TPE Tuner parameters:

- `tpe_args.constant_liar_type`:

  Type: `'best' | 'worst' | 'mean' | null`

  Default: `'best'`

TPE algorithm itself does not support parallel tuning. This parameter specifies how to optimize for `trial_concurrency` > 1. In general, `best` suit for small trial number and `worst` suit for large trial number.

- `tpe_args.n_startup_jobs`:

  Type: `int`

  Default: `20`

  The first *N* random trials are generated for warming up in Step 1. If the search space is large, this value should be increased.

- `tpe_args.n_ei_candidates`:

  Type: `int`

  Default: `24`

  n random candidates generated in Step 4.

- `tpe_args.linear_forgetting`:

  Type: `int`

  Default: `25`

  TPE lowers the weights of old trials. This parameter controls how many iterations it takes for a trial to start decay.

- `tpe_args.prior_weight`:

  Type: `float`

  Default: `1.0`

  Determines the weight of trial configuration in the history trial configurations.

- `tpe_args.gamma`:

  Type: `float`

  Default: `0.25`

  Controls how many trials are considered "good." Represents γ parameter from Step 2.

Note TPE Tuner configuration parameters mentioned above are only valid from NNI version 2.6. They will not work on previous versions.

TPE Tuner supports all search space types: choice, choice(nested), randint, uniform, quniform, loguniform, qloguniform, normal, qnormal, lognormal, and qlognormal.

We can see from Listing 3-7 how TPE Tuner performs optimizing holder_function.

(Full code is provided in the corresponding file: *ch3/tuners/tpe_tuner/run_experiment.py*.)

Setting the TPE Tuner for the Experiment:

Listing 3-7. TPE Tuner

```
experiment.config.tuner.name = 'TPE'
experiment.config.tuner.class_args['optimize_mode'] = 'maximize'
experiment.config.tuner.class_args['seed'] = 0
experiment.config.tuner.class_args['tpe_args'] = {
    'n_startup_jobs': 20,
    'gamma':          0.5
}
```

After the experiment is completed, we can analyze the progress of the search for TPE Tuner.

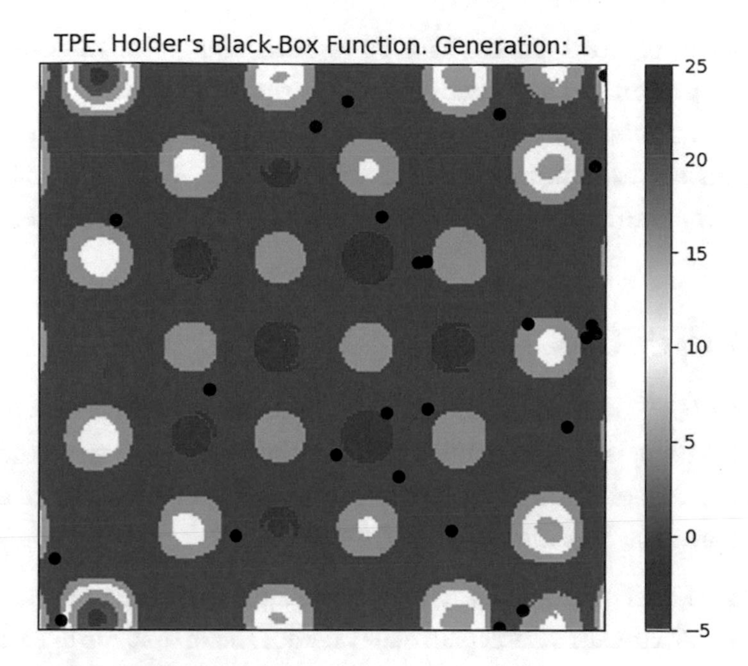

Figure 3-21. *TPE Tuner. Generation: 1*

Figure 3-21 shows that the distribution of points is more like a random scattering, which makes sense because, in the tuner setup, we specified `'n_startup_jobs': 20`, which means that the first 20 trials will be completely random.

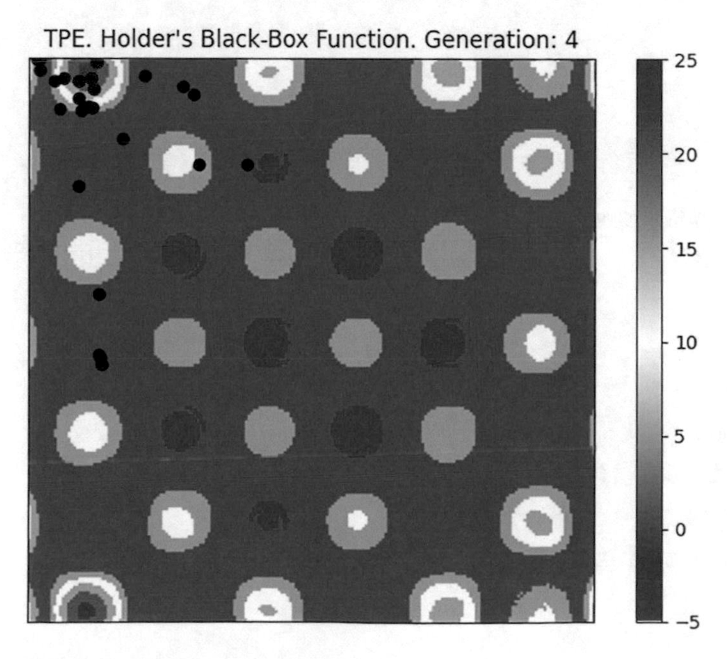

Figure 3-22. *TPE Tuner. Last generation*

But as we see in Figure 3-22, the TPE Tuner finds the global maxima of `holder_function`, thanks to a probabilistic exploration model.

The TPE Tuner is intuitively clear, based on a solid probability ground, and has a good balance between exploration and exploitation policy. Another nice feature is that it supports `choice(nested)` search type, which can be critical in some research.

Gaussian Process Tuner

Gaussian Process (GP) Tuner is another SMBO Tuner based on Multivariate Normal Distribution. This tuner is similar to the TPE Tuner but uses a Gaussian distribution to build the p(y|x) surrogate. A full description of this method lies outside of this book, but the reader can learn how this method works here:

- "Using Gaussian Processes to Optimize Expensive Functions": https://citeseerx.ist.psu.edu/viewdoc/download?doi=10.1.1.139.9315&rep=rep1&type=pdf
- "Gaussian Processes and Bayesian Optimization": www.cs.cornell.edu/courses/cs4787/2019sp/notes/lecture16.pdf

Here is an example of GP Tuner configuration:

```
# config.yml
tuner:
  name: GPTuner
  classArgs:
    optimize_mode: maximize
    utility: 'ei'
    kappa: 5.0
    xi: 0.0
    nu: 2.5
    alpha: 1e-6
    cold_start_num: 10
    selection_num_warm_up: 100000
    selection_num_starting_points: 250
```

The following is the description of GP Tuner parameters:

- `utility`:

  Type: `'ei' | 'ucb' | 'poi'`

  Default: `'ei'`

  The utility functions `ei`, `ucb`, and `poi` correspond to Expected Improvement, Upper Confidence Bound, and Probability of Improvement, respectively.

- `kappa`:

  Type: `float`

  Default: `5`

  Used by the `ucb` utility function. The bigger the `kappa` is, the more exploratory the tuner will be.

- `xi`:

  Type: `float`

  Default: `0`

  Used by the `ei` and `poi` utility functions. The bigger the `xi` is, the more exploratory the tuner will be.

- `nu`:

  Type: `float`

  Default: `2.5`

  Sets the Matern kernel. The smaller the nu is, the less smooth the approximated function will be.

- `alpha`:

  Type: `float`

  Default: `1e-6`

  Sets the Gaussian Process Regressor. Larger values correspond to an increased noise level in the observations.

- `cold_start_num`:

  Type: `int`

  Default: `10`

  Number of random explorations to perform before the Gaussian Process.

- `selection_num_warm_up`:

  Type: `int`

  Default: `1e5`

  Number of random points to evaluate when getting the point which maximizes the acquisition function.

- `selection_num_starting_points`:

  Type: `int`

  Default: `250`

  Number of times to run L-BFGS-B from a random starting point after the warm-up.

GP Tuner supports the following search space types: `choice`, `randint`, `uniform`, `quniform`, `loguniform`, and `qloguniform`.

GP Tuner suffers a lot from parallelization issues. If you run an experiment in concurrency mode (i.e., `trial_concurrency` > 1), multiple processes simultaneously decide on their next trial candidate based on the same historical data. Therefore, different processes are testing the same parameters at the same time. This is a big problem with all SMBO tuners. But TPE Tuner can get around this problem with the Constant Liar technique, while for GP Tuner, this problem remains serious. In Figure 3-23, we can see the Trial Metric panel for GP Tuner with `trial_concurrency = 8`. It shows that GP Tuner contains chunks of the same trials, which does not speed up the process of exploring the search space in any way.

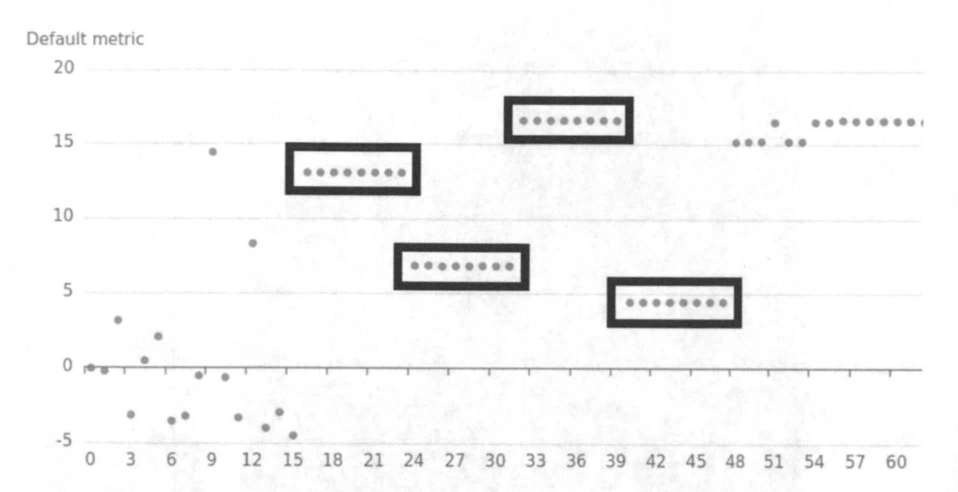

Figure 3-23. *GP Tuner. Concurrency issue*

Listing 3-8 implements GP Tuner for `holder_function` optimization task. (Full code is provided in the corresponding file: *ch3/tuners/gp_tuner/run_experiment.py*.)

Disabling concurrency:

Listing 3-8. GP Tuner

```
experiment.config.trial_concurrency = 1
```

Setting the GP Tuner for the Experiment:

```
experiment.config.tuner.name = 'GPTuner'
experiment.config.tuner.class_args['optimize_mode'] = 'maximize'
```

GP Tuner shows excellent results. Figure 3-24 shows the coordinates of all trials during the experiment. GP Tuner found both global maxima and evenly explored the entire search space.

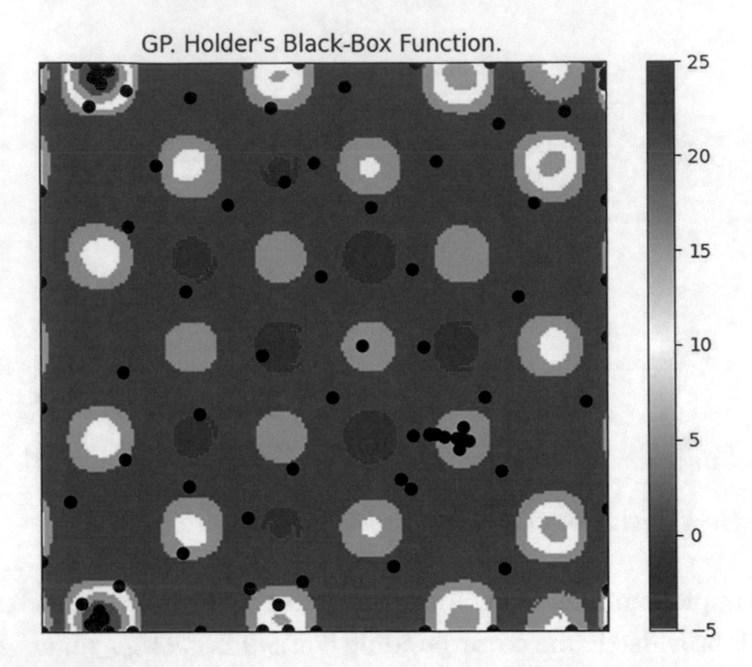

Figure 3-24. *GP Tuner. Holder's black-box function optimization*

GP Tuner suggests taking search space elements to find a suitable solution in a small number of black-box function evaluations. GP Tuner carries about the surrogate model instead of the black-box function itself, using the conjugate gradient method to find the highest expected improvement candidates. GP Tuner shows good exploratory behavior in testing out areas that seem promising under the current surrogate model.

Which Tuner to Choose?

In this section, we've put a lot of effort into learning about the different tuners, and the fair question might be: *Hey, so which tuner should I choose?* And the answer to this question is disappointing: according to the **No Free Lunch Theorem**, which we already considered in Chapter 1, ***there is no search algorithm (Tuner) that would have any advantages over other Tuners for an arbitrary search space***. Indeed, for an arbitrary search space, the expectation of the TPE Tuner does not exceed the Random Tuner expectation. So why do we need any Tuners at all if all of them are equal to Random Tuner on an arbitrary search space?!

And the answer is as follows: search spaces have a specific structure and dependencies that allow certain Tuners to outperform others for many kinds of problems. Therefore, if we know that we are optimizing a model for an image classification problem, this can give us some insight into the search space structure. Consequently, we can choose a Tuner that is more likely to show good results than others.

There is a separate area of research in which scientists arrange battles of search algorithms to determine the best one for a particular class of problems. Scientists use benchmarks to estimate the characteristics of the search algorithm. The benchmark algorithm evaluates the search algorithm several times for different search spaces. For example, the benchmark pseudo-code might look like this:

```
# search_algos: List of competing Search Algorithms
# problems: List of similar problems
# results: Map (Dict) of results

for algo in search_algos:
    for p in problems:
        metrics = algo(p)
        results[algo].append(metrics)

# Sort algorithms by results
```

In Table 3-1, I provide a sample benchmark result obtained using NNI https://nni.readthedocs.io/en/v2.7/hpo/hpo_benchmark_stats.html.

Table 3-1. *Average rankings for classification tasks*

Tuner Name	Average Ranking
GP Tuner	4.00
Evolution	4.22
Anneal	4.39
TPE	4.67
Random	5.33

Some benchmarks can last several days or even weeks. Therefore, it is always more convenient to borrow the results obtained and published after research. In many cases, you can execute a mini-research yourself, which will help determine the characteristics of the search space structure. In any case, understanding the deep learning model optimization problem and principles of the Search Tuner is very helpful in choosing the right strategy for solving the HPO problem.

Custom Tuner

Built-in tuners are suitable for most tasks. But there are situations when you need to add some custom logic to improve the quality of the HPO Experiment. Indeed, sometimes, we may know specific properties of the search space that the built-in tuner does not take into account. Also, the developer can implement their original idea and test it on real problems. For such cases, NNI allows you to implement a Custom Tuner. Custom Tuner can be used in an experiment and shared with other developers.

Tuner Internals

Each Tuner class should inherit `nni.tuner.Tuner` and implement the following methods: `__init__`, `update_search_space`, `generate_parameters`, `receive_trial_result`. Any Tuner can be implemented based on the self-describing sample presented in Listing 3-9.

Listing 3-9. Custom Tuner. ch3/tuners/custom_tuner/custom_tuner_sample.py

```
from nni.tuner import Tuner

class CustomTunerSample(Tuner):

    def __init__(self, some_arg) -> None:
        # YOUR CODE HERE #
        ...

    def update_search_space(self, search_space):
        """
        Tuners are advised to support updating search
        space at run-time. If a tuner can only set
```

```
        search space once before generating first
        hyper-parameters, it should explicitly document
        this behaviour. 'update_search_space' is called
        at the startup and when the search space is updated.
        """

        # YOUR CODE HERE #
        ...

    def generate_parameters(self, parameter_id, **kwargs):
        """
        This method will get called when the framework
        is about to launch a new trial. Each parameter_id
        should be linked to hyper-parameters returned by
        the Tuner. Returns hyper-parameters, a dict
        in most cases.
        """

        # YOUR CODE HERE #
        # Example: return {"dropout": 0.5, "act": "relu"}

        return {}

    def receive_trial_result(self, parameter_id, parameters, value,
    **kwargs):
        """
        This method is invoked when a trial reports
        its final result. Should be implemented
        if Tuner assumes 'memory', i.e.,
        Tuner is tracking previous Trials
        """
        # YOUR CODE HERE #
```

Tuner interacts with Experiment the following way:

1. Experiment calls `update_search_space` at the startup.
2. Experiment requests search space parameters for Trial calling `generate_parameters`.

3. Experiment returns Trial results to Tuner calling `recieve_trial_result`.

(Steps 2 and 3 are repeated until `max_trial_number` is reached or Experiment is stopped.)

Figure 3-25 shows Tuner–Experiment interaction as a sequence diagram.

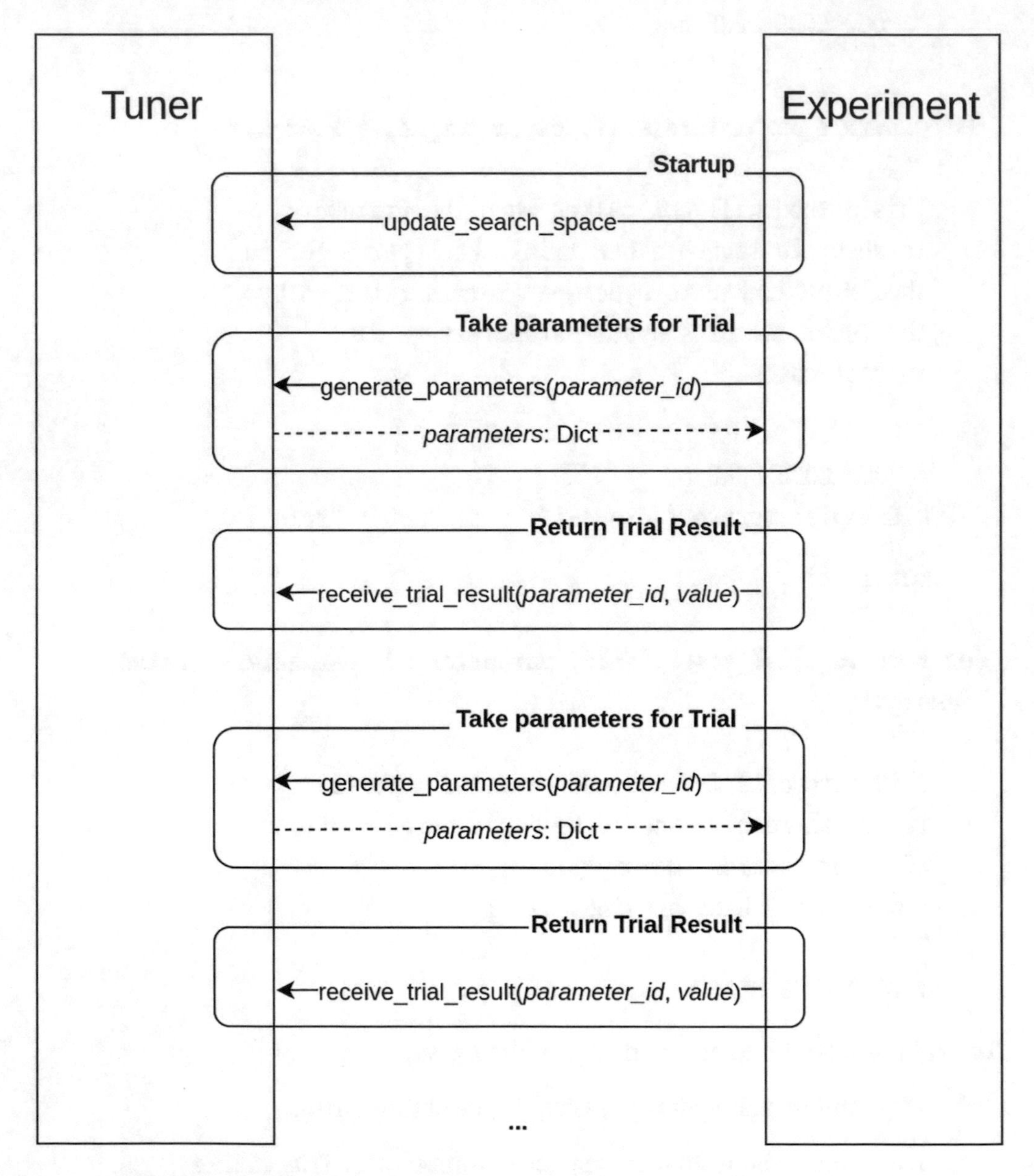

Figure 3-25. *Tuner–Experiment sequence diagram*

Custom Tuner is integrated into the experiment using the following config:

```
tuner:
  codeDirectory: <path_to_tuner_dir>
  className: <tuner_file_name>.<class_name>
```

Or it can be integrated into Python embedded experiment as follows:

```
from nni.experiment import CustomAlgorithmConfig

experiment.config.tuner = CustomAlgorithmConfig()
experiment.config.tuner.code_directory = 'path_to_tuner_dir'
experiment.config.tuner.class_name = 'tuner_file_name.class_name'
experiment.config.tuner.class_args = {'arg': 'value'}
```

New Evolution Custom Tuner

Let's try to develop our Custom Tuner. This Tuner will be based on the evolutionary concepts we examined exploring the Evolution Tuner. We'll call it `NewEvolutionTuner`. `NewEvolutionTuner` will initialize the population and act according to the following algorithm:

- Take the best individual: X_{best}
- Mutate the best individual in a random way: $mutate(X_{best}) \rightarrow Y$
- Replace the worst individual in the population X_{worst} with a mutant of the best individual Y: $X_{worst} \leftarrow Y$

Figure 3-26 demonstrates the search approach of `NewEvolutionTuner`:

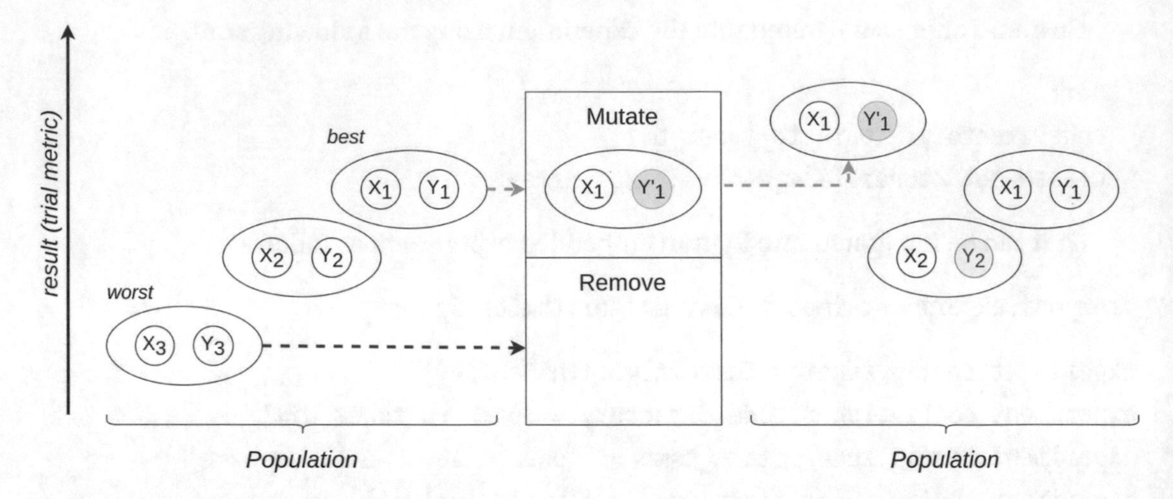

Figure 3-26. *New Evolution Tuner*

`NewEvolutionTuner` implements a greedy "hill-climbing" approach. The Tuner takes the best individual (the highest one) and mutates it (hoping that the new individual will climb higher). Listing 3-10 use `NewEvolutionTuner` to find the maxima of Ackley's function.

Listing 3-10. Ackley's function. ch3/bbf/ackley.py

```
from numpy import exp, sqrt, cos, pi, e
from ch3.bbf.utils import noise

def ackley_function(x, y):
    """
    Ackley's function
    """
    z = 20.0 * exp(-0.2 * sqrt(0.5 * (x**2 + y**2))) -\
        exp(0.5 * (cos(2 * pi * x) + cos(2 * pi * y))) + e + 20
    r = z + noise(x, y, scale = 4)
    return r
```

And here is Ackley's function visualization:

```
from ch3.bbf.utils import scatter_plot
if __name__ == '__main__':
    scatter_plot(ackley_function, [-10, 10], [-10, 10])
```

We can see in Figure 3-27 the surface of Ackley's function. It has one highest hill and several smaller hills nearby.

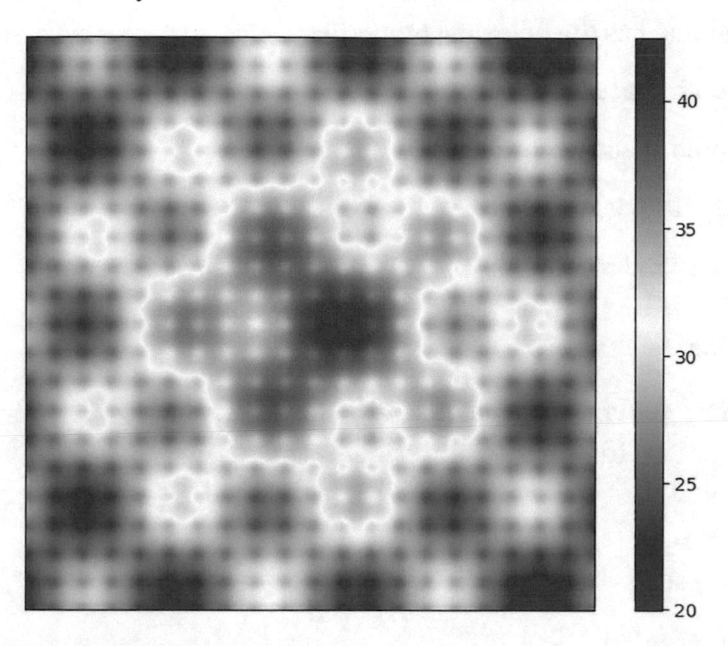

Figure 3-27. *Ackley's function*

Before we check how `NewEvolutionTuner` solves the problem of finding the maxima of Ackley's function, we need to implement it in Listing 3-11.

Importing necessary modules:

Listing 3-11. NewEvolutionTuner. ch3/tuners/custom_tuner/evolution_tuner.py

```
import random
import numpy as np
from nni.tuner import Tuner
from nni.utils import (
    OptimizeMode, extract_scalar_reward,
    json2space, json2parameter,
)
```

Population consists of individuals. Each individual has DNA that represents a search space parameter. Also, individual has a result field that contains Trial result. New Evolution individual has the following properties:

- x: Search space *x* coordinate
- y: Search space *y* coordinate
- param_id: Trial number
- result: Trial result

```
class Individual:

    def __init__(self, x, y, param_id = None) -> None:
        self.param_id = param_id
        self.x = x
        self.y = y
        self.result = None

    def to_dict(self):
        return {'x': self.x, 'y': self.y}
```

Population class is a wrapper that manipulates all individuals as a whole:

```
class Population:
```

All individuals are stored in individuals property:

```
    def __init__(self) -> None:
        self.individuals = []

    def add(self, ind):
        self.individuals.append(ind)
```

Then, we need to add a method that will return an individual by its param_id:

```
    def get_by_param_id(self, param_id):
        for ind in self.individuals:
            if ind.param_id == param_id:
                return ind
        return None
```

At the start of the experiment, the Tuner will create *N* individuals who will not have a trial number. We will call these individuals virgins. `get_first_virgin` method returns the first virgin found in population:

```
def get_first_virgin(self):
    for ind in self.individuals:
        if ind.param_id is None:
            return ind
    return None
```

The following method - `get_population_with_result` - returns all individuals that already received a trial result:

```
def get_population_with_result(self):
    population_with_result = [ind for ind in self.individuals if
    ind.result is not None]
    return population_with_result
```

The next method returns the best individual from the whole population, that is, an individual that has the highest `result`:

```
def get_best_individual(self):
    sorted_population = sorted(self.get_population_with_result(),
    key = lambda ind: ind.result)
    return sorted_population[-1]
```

And here, we come to our primary evolution method `replace_worst`, which will develop the population:

- We take the best individual from the population.
- Mutate the best individual.
- Add the mutant of the best individual to the population instead of the worst one.

```
def replace_worst(self, param_id):
    population_with_result = self.get_population_with_result()
    sorted_population = sorted(population_with_result, key = lambda
    ind: ind.result)
    worst = sorted_population[0]
```

```
        self.individuals.remove(worst)
        best = self.get_best_individual()
        x = round(best.x + random.gauss(0, 1), 2)
        y = round(best.y + random.gauss(0, 1), 2)
        mutant = Individual(x, y, param_id)
        self.individuals.append(mutant)
        return mutant
```

We can start implementing the tuner after defining the Individual and Population classes:

```
class NewEvolutionTuner(Tuner):
```

The Tuner has two parameters, optimize_mode and population size:

```
    def __init__(self, optimize_mode = "maximize", population_size = 16)
    -> None:

        self.optimize_mode = OptimizeMode(optimize_mode)
        self.population_size = population_size
```

Next, Tuner initializes properties related to the search space it is working with:

```
        self.search_space_json = None
        self.random_state = None
        self.population = Population()
        self.space = None
```

When the Tuner starts, the update_search_space method is invoked. It generates the Population of Random Individuals:

```
    def update_search_space(self, search_space):
        self.search_space_json = search_space
        self.space = json2space(self.search_space_json)
        self.random_state = np.random.RandomState()

        # Population of Random Individuals is generated
        is_rand = dict()
        for item in self.space:
            is_rand[item] = True
```

```
    for _ in range(self.population_size):
        params = json2parameter(self.search_space_json, is_rand, self.
        random_state)
        ind = Individual(params['x'], params['y'])
        self.population.add(ind)
```

Experiment calls generate_parameters method to get new parameters for the subsequent Trial. Initially, we have a set of individuals generated at the startup and not passed to the Tuner (virgins). We take virgins and pass them to Tuner one by one. When no virgins are left, we are generating new individuals.

```
def generate_parameters(self, parameter_id, **kwargs):
    virgin = self.population.get_first_virgin()
    if virgin:
        virgin.param_id = parameter_id
        return virgin.to_dict()
    else:
        mutant = self.population.replace_worst(parameter_id)
        return mutant.to_dict()
```

When the Experiment returns the Trial's result, we save it to an individual object:

```
def receive_trial_result(self, parameter_id, parameters, value,
**kwargs):
    reward = extract_scalar_reward(value)
    ind = self.population.get_by_param_id(parameter_id)
    ind.result = reward
```

Well, our NewEvolutionTuner is ready for action! We can launch the experiment using the following config file shown in Listing 3-12.

Listing 3-12. NewEvolutionTuner Experiment configuration. ch3/tuners/custom_tuner/config.yml

```
searchSpace:
  x:
    _type: "quniform"
    _value: [-10, 10, 0.01]
```

```
  y:
    _type: "quniform"
    _value: [-10, 10, 0.01]

trialConcurrency: 4
trialCodeDirectory: .
trialCommand: python3 trial.py

tuner:
  codeDirectory: .
  className: evolution_tuner.NewEvolutionTuner

trainingService:
  platform: local
```

The experiment with custom `NewEvolutionTuner` can also be run in Python embedded mode as it is implemented in Listing 3-13.

(Full code is provided in the corresponding file: *ch3/tuners/custom_tuner/run_experiment.py*.)

Listing 3-13. Python embedded experiment

```
experiment.config.tuner = CustomAlgorithmConfig()
experiment.config.tuner.code_directory = Path(__file__).parent
experiment.config.tuner.class_name = 'evolution_tuner.NewEvolutionTuner'
experiment.config.tuner.class_args = {'population_size': 8}
```

Let's launch the Experiment and analyze its results:

```
$ python3 ch3/tuners/custom_tuner/run_experiment.py
```

Figure 3-28 shows the locations visited by the `NewEvolutionTuner` population. We see that the population has found the global function's maxima and began to explore it.

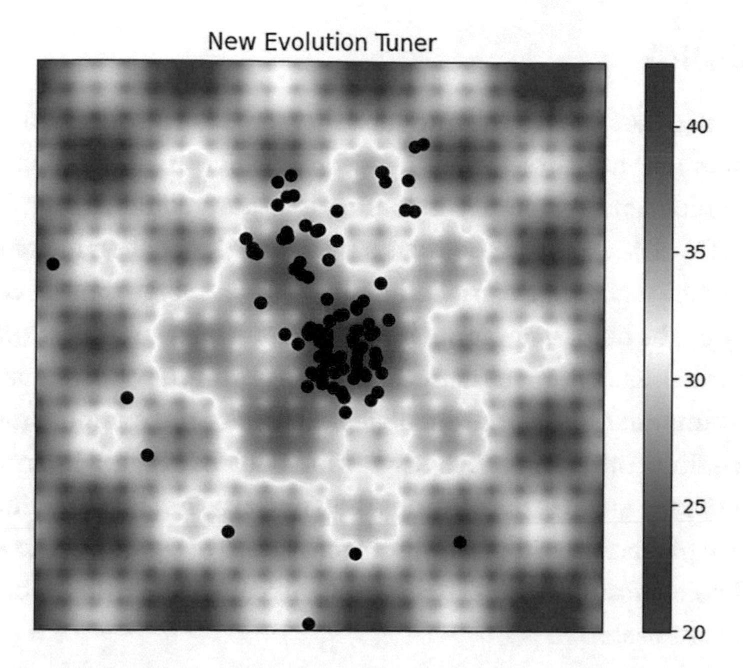

Figure 3-28. *Locations visited by population of NewEvolutionTuner*

Trial Metric panel in Figure 3-29 demonstrates the “hill-climbing” approach in action. The main problem with this approach is that it quickly finds a local maxima and stops exploring other search space areas.

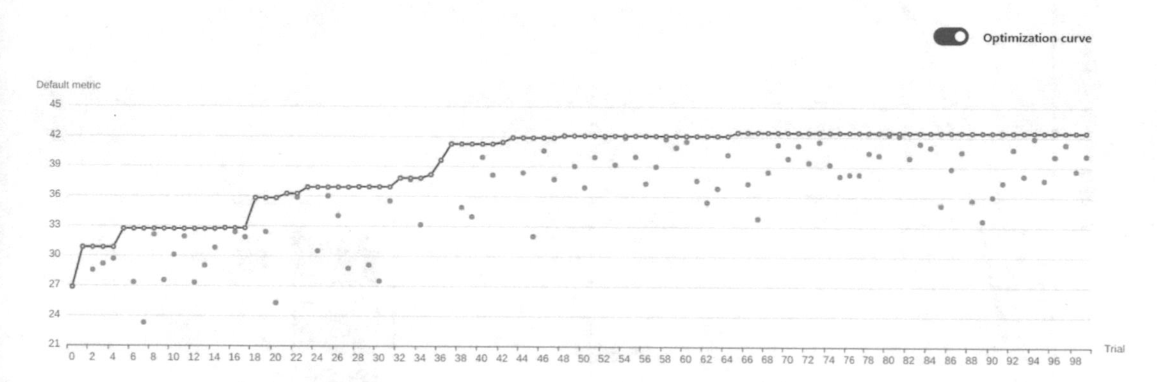

Figure 3-29. *Trial Metric panel*

Of course, developing a custom Tuner is not always an easy task. Still, the ability to implement your search algorithm and integrate it into the HPO process can greatly increase the experiment results. In this section, we have provided an example of how you can do this. If necessary, you can implement your ideas based on the illustration given in this section.

Early Stopping

Some parameters in the search space produce very low Trial results. And this is normal because the Tuner may not always know in advance which areas of the search space to explore and which not to explore. Tuner often tries parameters that give very low results. The Trial itself is expensive because a lot of time is spent on it. For example, training a neural network with a complex architecture on a large dataset can take hours. And it would be helpful not to spend a lot of time on trials that show poor results in their execution. NNI uses Early Stopping algorithms to solve that issue. Early Stopping algorithms analyze the intermediate trial results and compare them with the intermediate results of other trials. If the algorithm decides that the intermediate results of the current Trial are too low, then it stops the Trial so as not to waste time on it.

Figure 3-30 explains Early Stopping approach. Trial 3 early stopped at step N because intermediate results of this Trial were significantly worse than the other trials' intermediate results at step N.

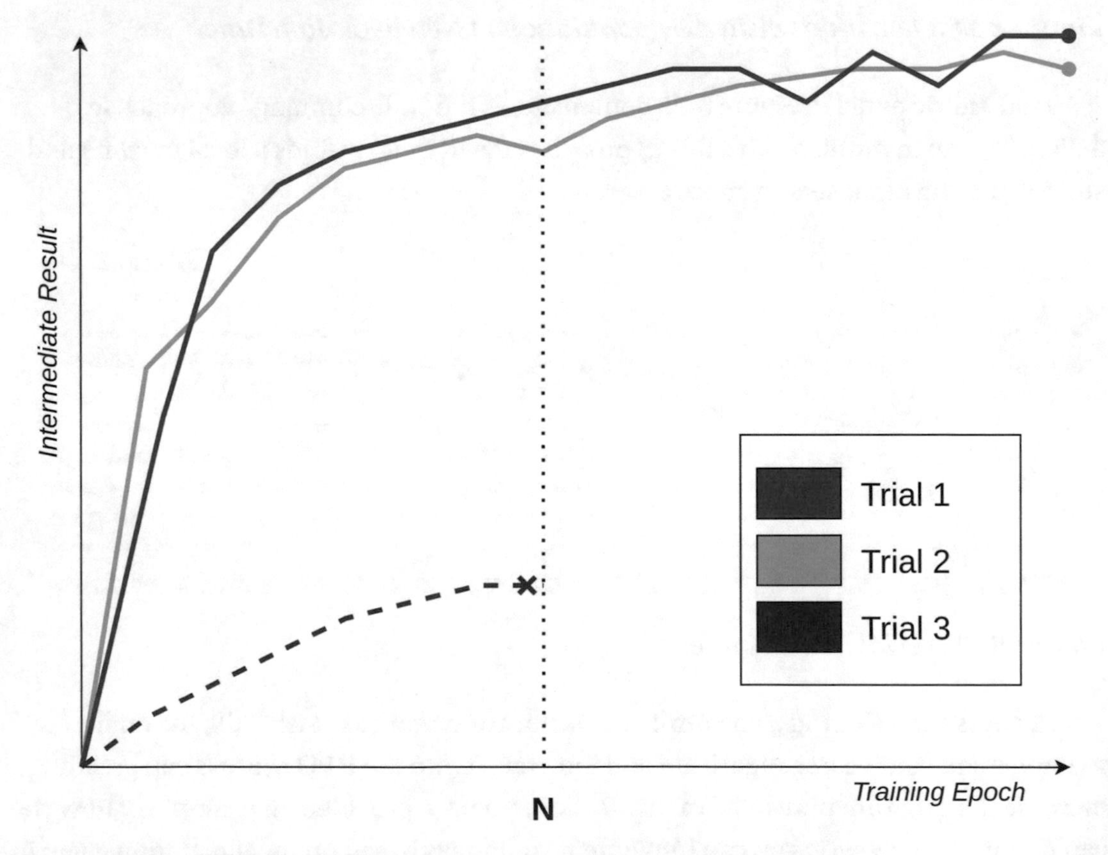

Figure 3-30. *HPO Early Stopping*

The deep learning training algorithms also have Early Stop policies. Training Early Stopping policy stops model training when the model starts to degrade or there is no improvement for a long time. Do not confuse Training Early Stopping with HPO Early Stopping. They are not related in any way. Indeed, take a look at Figure 3-31. Training progress is good, and there is no reason to stop the training. But if we compare the training process with other trials, it is apparent that it is much worse, and the HPO Early Stopping algorithm can stop this Trial.

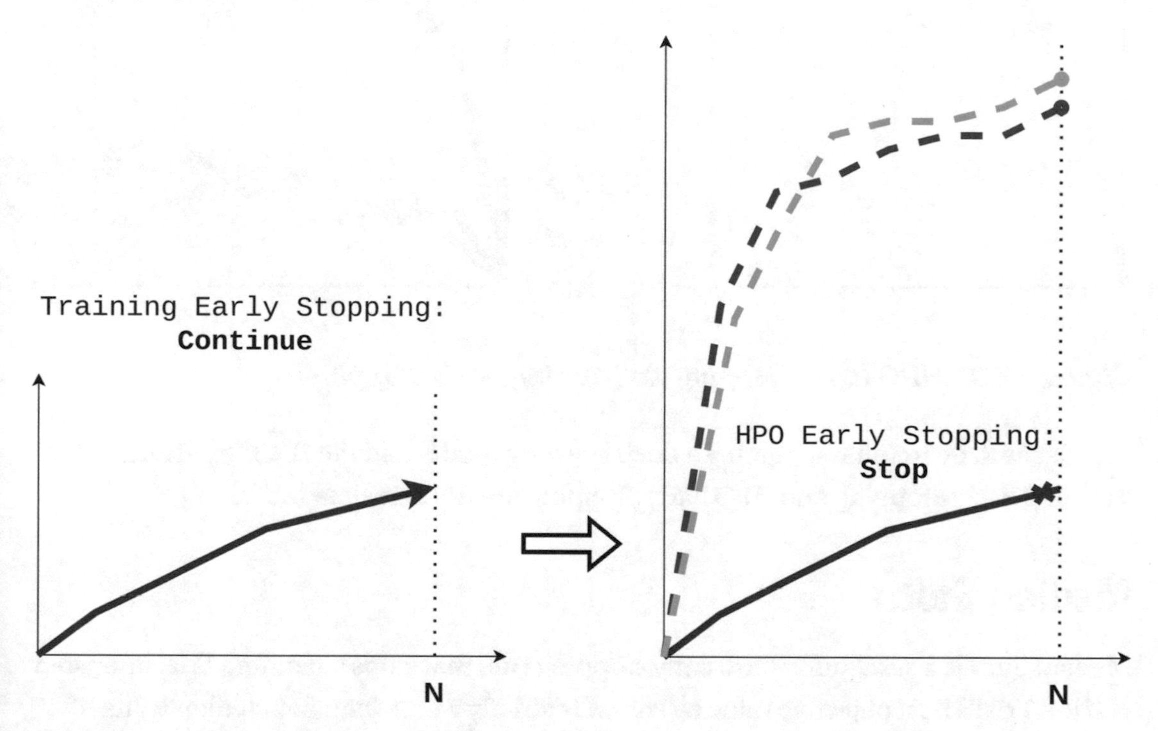

Figure 3-31. *HPO Early Stopping vs. Training Early Stopping*

And Figure 3-32 demonstrates the opposite situation. The training process begins to degrade, and the Training Early Stopping algorithm terminates the training process. In contrast, the HPO Early Stopping algorithm may consider the current Trail very promising because its intermediate results are significantly superior compared to other trials.

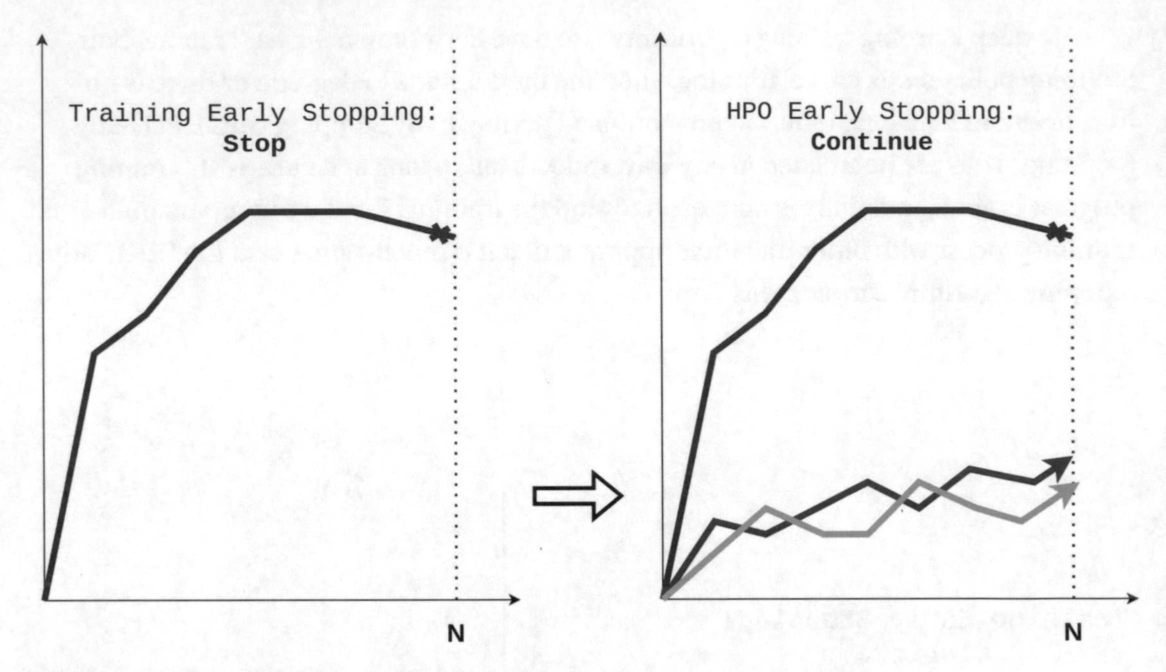

Figure 3-32. *HPO Early Stopping vs. Training Early Stopping*

Please keep in mind designing a deep learning model and the HPO Experiment that Training Early Stopping and HPO Early Stopping are not correlated.

Median Stop

Median Stop is a straightforward, early stopping rule that stops a pending Trial after step N if the Trial's best objective value by step N is strictly worse than the median value of the running averages of all completed trials' objectives reported up to step N.

Median Stop algorithm can be implemented in Experiment with the following experiment configuration:

```
assessor:
  name: Medianstop
  classArgs:
    # number of warm up steps
    start_step: 10
```

Let's look at a synthetic problem to see the Median Stop algorithm in action. Say, we have the following identity function, f: x → x, with the training progress containing 100 epochs (steps) and expressed by the following rule: $\frac{x}{10}\sqrt{epoch}\ + r$, where r is a random variable on (-1, 1). We can characterize the function f as an identity function with parabolic training progress. Listing 3-14 contains the implementation of function f.

Listing 3-14. Identity function with parabolic training progress. ch3/early_stop/medianstop/model.py

```
import random

def identity_with_parabolic_training(x):
    history = [
        max(round(x / 10, 2) * pow(h, .5) + random.uniform(-3, 3), 0)
        for h in range(1, 101)
    ]
    return x, history
```

Let's visualize the training process of the following set of functions: f(0), f(10), f(20), ..., f(100).

```
if __name__ == '__main__':
    import matplotlib.pyplot as plt

    for x in range(0, 101, 10):
        final, history = identity_with_parabolic_training(x)
        plt.plot(history, label = str(x))

    plt.ylabel('Intermediate Result')
    plt.xlabel('Epochs')
    plt.legend()
    plt.show()
```

Figure 3-33 shows various training curves. This plot illustrates that the lower training curves are unpromising. Unpromising training can be stopped in advance according to the Early Stopping algorithm.

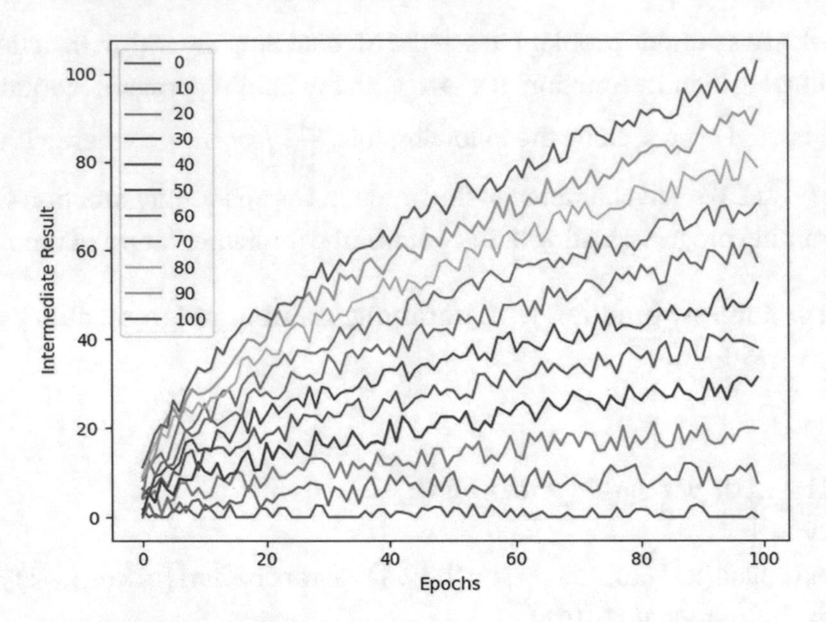

Figure 3-33. *Parabolic training*

Let's launch an experiment that will use the Median Stop algorithm. The trial script is defined in Listing 3-15.

Trial header with imported modules:

Listing 3-15. Median Stop Trial. ch3/early_stop/medianstop/trial.py

```
import os
import sys
from time import sleep
import nni

# For NNI use relative import for user-defined modules
SCRIPT_DIR = os.path.dirname(os.path.abspath(__file__)) + '/../../..'
sys.path.append(SCRIPT_DIR)
```

Executing Trial:

```
from ch3.early_stop.medianstop.model import identity_with_parabolic_training

if __name__ == '__main__':
    params = nni.get_next_parameter()
    x = params['x']
```

```
    final, history = identity_with_parabolic_training(x)
    for h in history:
        sleep(.1)
        nni.report_intermediate_result(h)

    nni.report_final_result(final)
```

Listing 3-16 defines the Experiment with Median Stop algorithm.

Listing 3-16. Experiment with Median Stop Algorithm. ch3/early_stop/medianstop/config.yml

```
searchSpace:
  x:
    _type: quniform
    _value: [1, 100, 0.1]

maxTrialNumber: 100
trialConcurrency: 8
trialCodeDirectory: .
trialCommand: python3 trial.py

tuner:
  name: Random

assessor:
  name: Medianstop
  classArgs:
    # number of warm up steps
    start_step: 10

trainingService:
  platform: local
```

Now we are ready to run the Experiment:

```
nnictl create --config ch3/early_stop/medianstop/config.yml
```

After the experiment is completed, we can observe in Figure 3-34 that many trials have the `EARLY_STOPPED` status, as expected.

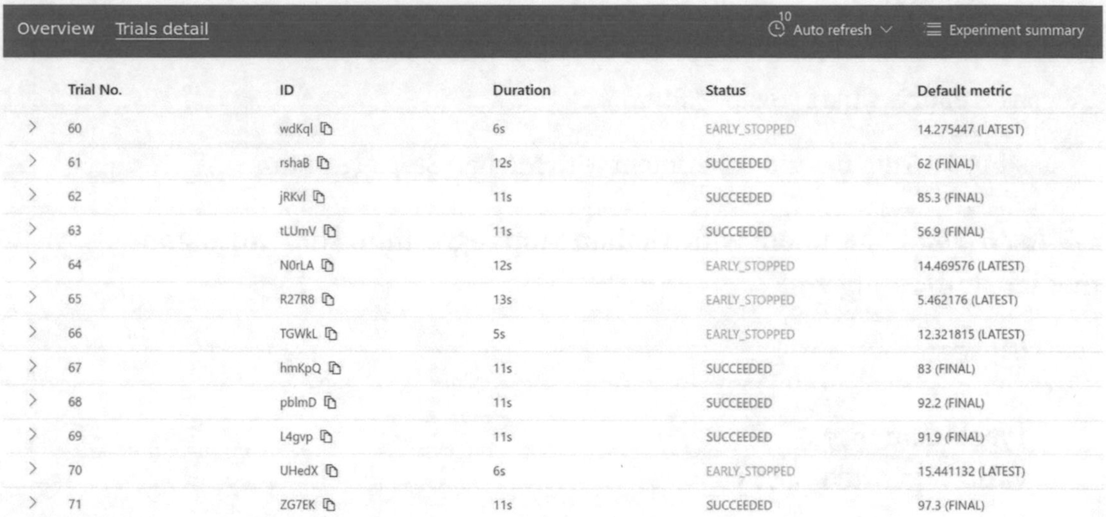

Overview Trials detail 10 Auto refresh Experiment summary

Trial No.	ID	Duration	Status	Default metric
60	wdKql	6s	EARLY_STOPPED	14.275447 (LATEST)
61	rshaB	12s	SUCCEEDED	62 (FINAL)
62	jRKvl	11s	SUCCEEDED	85.3 (FINAL)
63	tLUmV	11s	SUCCEEDED	56.9 (FINAL)
64	N0rLA	12s	EARLY_STOPPED	14.469576 (LATEST)
65	R27R8	13s	EARLY_STOPPED	5.462176 (LATEST)
66	TGWkL	5s	EARLY_STOPPED	12.321815 (LATEST)
67	hmKpQ	11s	SUCCEEDED	83 (FINAL)
68	pblmD	11s	SUCCEEDED	92.2 (FINAL)
69	L4gvp	11s	SUCCEEDED	91.9 (FINAL)
70	UHedX	6s	EARLY_STOPPED	15.441132 (LATEST)
71	ZG7EK	11s	SUCCEEDED	97.3 (FINAL)

Figure 3-34. *Trials detail panel*

Curve Fitting

Curve Fitting Assessor is an LPA (learning, predicting, assessing) algorithm that stops a pending Trial at step *N* if the prediction of the final epoch's performance is worse than the best final performance in the trial history. Curve Fitting Assessor makes a prediction about the final result of the Trial's training and compares it with the completed ones. This algorithm treats the Early Stopping task as a time series forecasting problem. If the training prediction is pessimistic, then the algorithm stops the trial. Figure 3-35 explains the Curve Fitting Early Stopping approach.

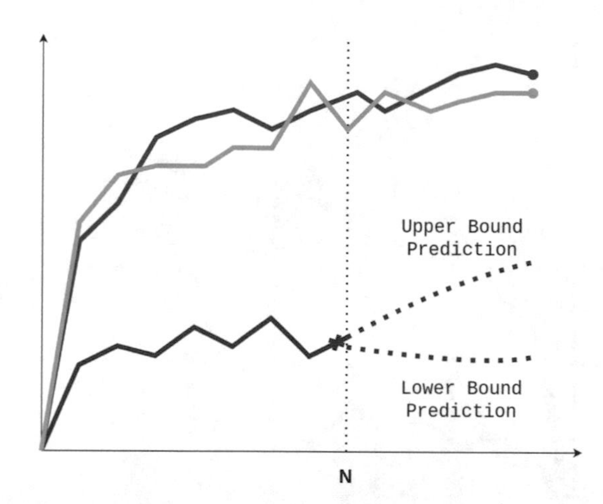

Figure 3-35. *Curve Fitting prediction*

Curve Fitting algorithm can be implemented in Experiment with the following experiment configuration:

```
assessor:
  name: Curvefitting
  classArgs:
    epoch_num: 20
    start_step: 6
    threshold: 0.95
    gap: 1
```

We will not study the principles of the Curve Fitting Early Stopping algorithm in this book. You can refer to the official documentation (`https://nni.readthedocs.io/en/v2.7/reference/hpo.html#nni.algorithms.hpo.curvefitting_assessor.CurvefittingAssessor`) or review a paper dedicated to "Speeding up Automatic Hyperparameter Optimization of Deep Neural Networks by Extrapolation of Learning Curves" (`https://ml.informatik.uni-freiburg.de/wp-content/uploads/papers/15-IJCAI-Extrapolation_of_Learning_Curves.pdf`).

Risk to Stop a Good Trial

The Early Stopping algorithm can significantly speed up the completion of an experiment and save computational resources. But there is always a risk of stopping a good trial too early, which would probably mean rejecting very good parameters in the search space. Look at Figure 3-36, early stopped trial could show good performance.

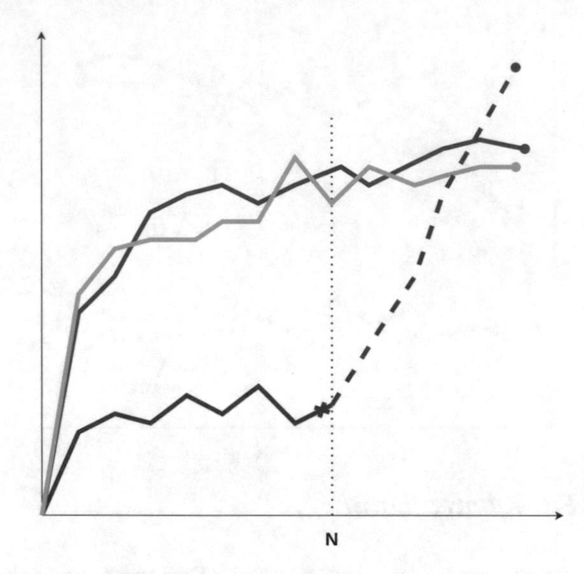

***Figure 3-36.** Good trial early stopped*

However, there is a very tiny chance of stopping a trial that can give good results too early. Usually, the training curves of deep learning models behave similarly, so if the trials' intermediate results were significantly worse than those of other trials, you probably should not expect anything good, but complete the trial and move on to the next one.

Searching for Optimal Functional Pipeline and Classical AutoML

Libraries: Scikit-learn

Let's go back to the problem we studied in the "From LeNet to AlexNet" section of Chapter 2. In this task, we built a functional pipeline that would optimally solve the image classification problem. Indeed, each layer of deep learning is a particular functional operator, and we tried to find the best pipeline of these operators using the HPO approach. Figure 3-37 presents neural network architecture as a functional pipeline.

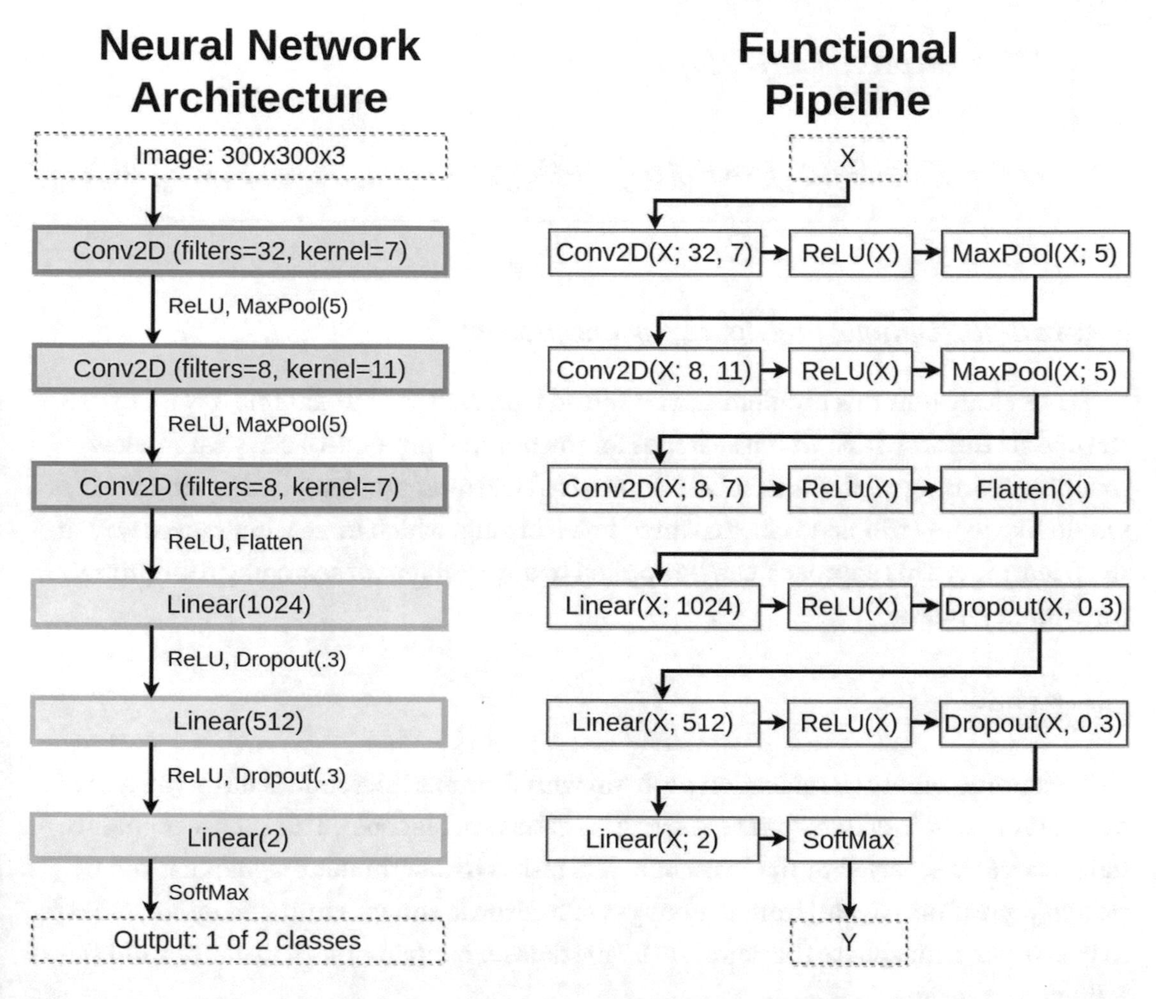

***Figure 3-37.** Neural network architecture as functional pipeline*

Let's convert the problem into a more strict mathematical language. We need to find a model M that consists of compositions of functions $F_i \in \{F\}$ and maximizes the value $L(M, D)$, where L evaluates the performance of model M on dataset D. *Figure 3-38 formulates the optimal functional pipeline problem:*

$$\begin{cases} M^* = \underset{M \in \{F_i\}}{\arg\max}(\, L(M, D)\,) \\ M^* = X \rightarrow \underbrace{F_1(p_1) \rightarrow F_2(p_2) \rightarrow F_3(p_3) \rightarrow \;\ldots\; \rightarrow F_n(p_n)}_{n} \rightarrow Y \end{cases}$$

Figure 3-38. *Optimal functional pipeline problem*

Let's study how this problem can be solved with NNI. As an example, I want to use the classic AutoML task, which searches for the optimal pipeline of classical shallow machine learning methods to solve a supervised learning problem. In this section, I would like to pay tribute to classical machine learning, which increasingly gives way to deep learning. This approach can be applied to any problem of searching the optimal functional pipeline.

Problem

Let's examine binary classification problem with Gamma Telescope Dataset (`https://archive.ics.uci.edu/ml/datasets/magic+gamma+telescope`). This dataset contains data of signals received by the telescope. The task is to discriminate signals caused by primary **gammas** (signal) from the images of **hadronic** showers initiated by cosmic rays in the upper atmosphere (background). This dataset contains 19020 instances and the following columns:

1. `fLength`: Type: real. Major axis of ellipse (mm)
2. `fWidth`: Type: real. Minor axis of ellipse (mm])
3. `fSize`: Type: real. 10-log of sum of content of all pixels (photons)
4. `fConc`: Type: real. Ratio of sum of two highest pixels over `fSize` (ratio)
5. `fConc1`: Type: real. Ratio of highest pixel over `fSize` (ratio)
6. `fAsym`: Type: real. Distance from highest pixel to center, projected onto major axis (mm)

7. `fM3Long`: Type: real. Third root of third moment along major axis (mm)

8. `fM3Trans`: Type: real. Third root of third moment along minor axis (mm)

9. `fAlpha`: Type: real. Angle of major axis with vector to origin (deg)

10. `fDist`: Type: real. Distance from origin to center of ellipse (mm)

11. `class`: Values: `'g'`, `'h'`. Gamma (signal), Hadron (background). g = gamma (signal): 12332. h = hadron (background): 6688

Perhaps the reader understands something in these physical data, but I do not understand anything about them. But that's exactly what we need machine learning for – to find patterns and dependencies where we cannot see them ourselves.

The dataset is located in ch3/ml_pipeline/data/magic04.data and is converted to a supervised learning problem in Listing 3-17.

Importing modules:

Listing 3-17. Gamma Telescope Dataset. ch3/ml_pipeline/utils.py

```
import os
from sklearn.model_selection import train_test_split
import pandas as pd
import numpy as np

def telescope_dataset():
```

Loading dataset from the file:

```
    cd = os.path.dirname(os.path.abspath(__file__))
    telescope_df = pd.read_csv(f'{cd}/data/magic04.data')
```

Dropping na values:

```
    telescope_df.dropna(inplace = True)
```

Setting column names:

```
    telescope_df.columns = [
        'fLength', 'fWidth', 'fSize', 'fConc', 'fConcl',
        'fAsym', 'fM3Long', 'fM3Trans', 'fAlpha', 'fDist', 'class']
```

Shuffling dataset:

```
telescope_df = telescope_df.iloc[np.random.
permutation(len(telescope_df))]
telescope_df.reset_index(drop = True, inplace = True)
```

Class labeling:

```
telescope_df['class'] = telescope_df['class'].map({'g': 0, 'h': 1})
y = telescope_df['class'].values
```

Splitting dataset on train and test datasets:

```
train_ind, test_ind = train_test_split(
    telescope_df.index,
    stratify = y,
    train_size = 0.8,
    test_size = 0.2
)

X_train = telescope_df.drop('class', axis = 1).loc[train_ind].values
X_test = telescope_df.drop('class', axis = 1).loc[test_ind].values

y_train = telescope_df.loc[train_ind, 'class'].values
y_test = telescope_df.loc[test_ind, 'class'].values

return X_train, y_train, X_test, y_test
```

Since we have identified the problem and prepared the dataset, we can begin to determine the machine learning methods that our model will consist of.

Operators

Now let's define the functions that will make up the functional pipeline. We will call them Operators in the machine learning context. Machine learning operators for a classification problem can be separated into three types:

- **Selectors**: Choose the most significant features (columns) from the dataset, removing dependent features. Usually reduces the input size. The selected features remain unchanged.

- **Transformers**: Convert input according to some mathematical function without changing the input size.
- **Classifiers**: Solve the classification problem itself.

Figure 3-39 shows a machine learning pipeline: selector, transformer, and classifier.

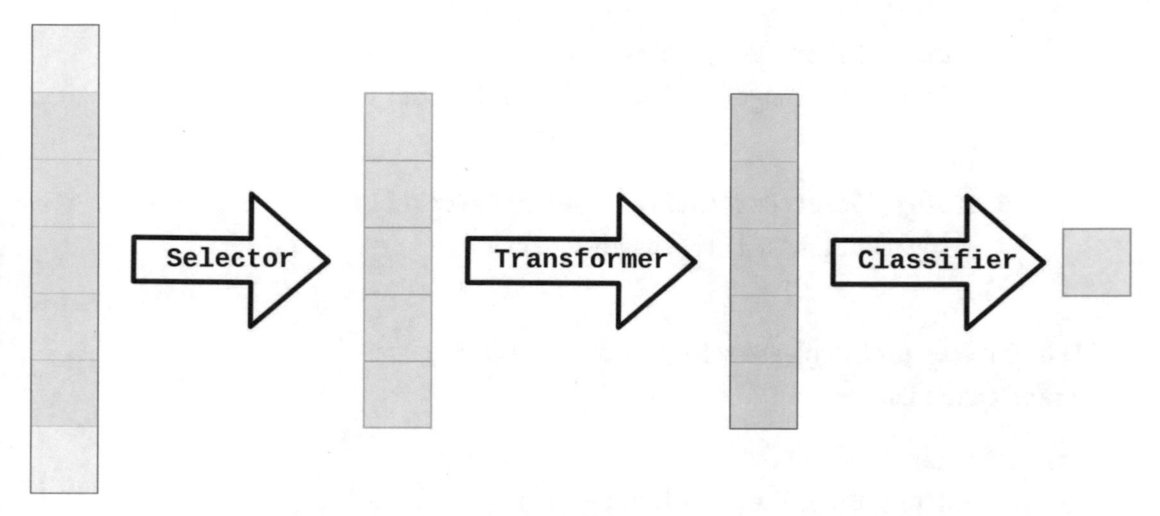

***Figure 3-39.** Machine learning operators: selector, transformer, and classifier*

Each operator has its own parameters, such as `max_depth` in the `DecisionTreeClassifier`. Listing 3-18 creates an operator space that will be used for the AutoML pipeline.

(Full code is provided in the corresponding file: *ch3/ml_pipeline/operator.py*.)

Each operator is an object that has a name, implementation, and parameters:

Listing 3-18. Operator space

```
class Operator:

    def __init__(self, name, clz, params = None):
        if params is None:
            params = {}
        self.name = name
        self.clz = clz
        self.params = params
```

Next, we define the operator space:

```
class OperatorSpace:
```

Selector list:

```
selectors = [
    Operator('SelectFwe', SelectFwe, {
        'alpha': arange(0, 0.05, 0.001).tolist()
    }),

    Operator('SelectPercentile', SelectPercentile, {
        'percentile': list(range(1, 100))
    }),
```

(For full selector list, please refer to the source code.)

Transformer list:

```
transformers = [
    Operator('Binarizer', Binarizer, {
        'threshold': arange(0.0, 1.01, 0.05).tolist()
    }),

    Operator('FastICA', FastICA, {
        'tol': arange(0.0, 1.01, 0.05).tolist()
    }),
```

(For full transformer list, please refer to the source code.)

Classifier list:

```
classifiers = [
    Operator('GaussianNB', GaussianNB),

    Operator('BernoulliNB', BernoulliNB, {
        'alpha': [0.01, 0.1, 1, 10]
    }),
```

(For full classifier list, please refer to the source code.)

Finally, we add an additional auxiliary method `get_operator_by_name`, which returns the operator by its name:

```
@classmethod
def get_operator_by_name(cls, name):

    operators = cls.selectors + cls.transformers + cls.classifiers
    for o in operators:
        if o.name == name:
            return o
    return None
```

Search Space

Let's define search space for the classifier. We assume that the classifier's operator pipeline will have

- **Selectors**: From 0 to 1 (selector is advised but not required)
- **Transformers**: From 0 to 3 (selector is advised but not required)
- **Classifier**: 1 (is required)

Therefore, a pipeline can have from one to five operators. Please take a look at Figure 3-40.

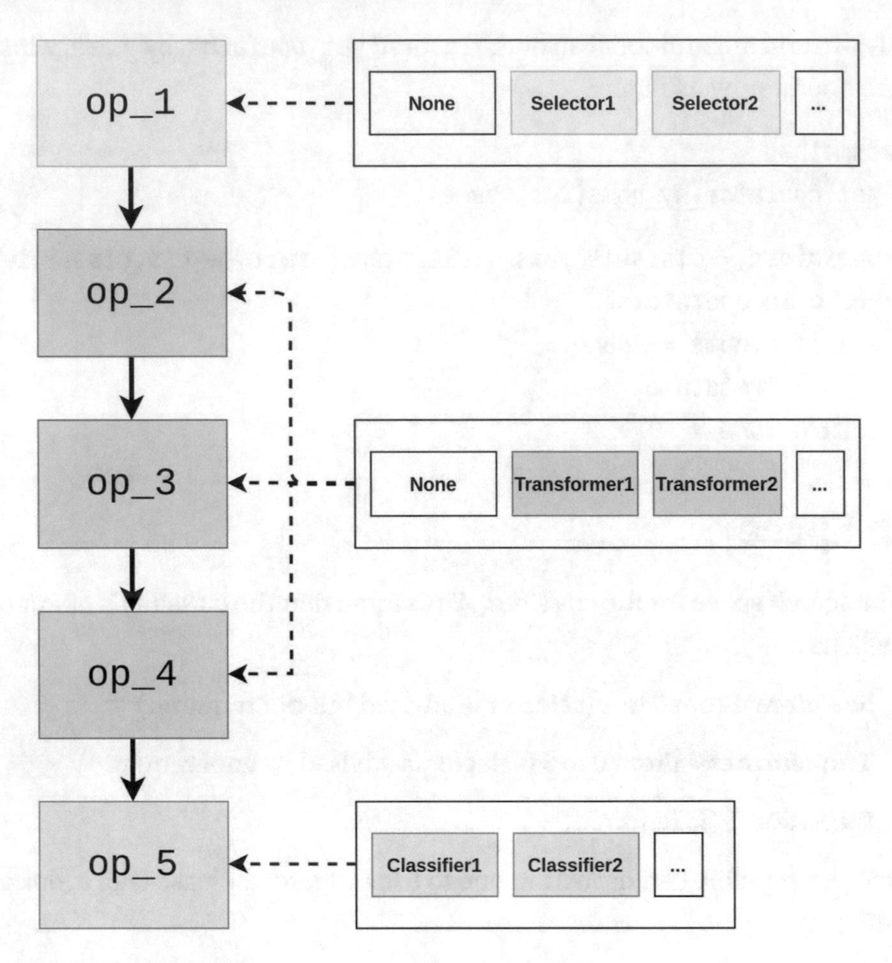

Figure 3-40. *Pipeline cells*

The pipeline has five cells. Each cell can be filled with some value from the corresponding operator space. The cell can be empty if `none` value is selected. For example, the following pipeline might be selected: `Selector3` → `none` → `Transformer1` → `none` → `Classifier2`, which is equal to `Selector3` → `Transformer1` → `Classifier2`. This search space definition is huge and challenging to construct manually, so we will add a special class in Listing 3-19 that will create an operator search space definition according to the NNI specification.

Listing 3-19. Search space. ch3/ml_pipeline/search_space.py

```
from ch3.ml_pipeline.operator import OperatorSpace

class SearchSpace:
```

Each cell has an operator type that can be filled by `selector`, `transformer`, and `classifier`. The `operator_search_space` method creates a search space for each cell type.

```
    @classmethod
    def operator_search_space(cls, operator_type):
        """
        Search space for operator by `operator_type`
        """
        ss = []
        operators = []

        if operator_type == 'selector':
            # Selectors are not required in Pipeline
            ss.append({'_name': 'none'})
            operators = OperatorSpace.selectors
        elif operator_type == 'transformer':
            # Transformers are not required in Pipeline
            ss.append({'_name': 'none'})
            operators = OperatorSpace.transformers
        elif operator_type == 'classifier':
            operators = OperatorSpace.classifiers

        for o in operators:
            row = {'_name': o.name}
            for p_name, values in o.params.items():
                row[p_name] = {"_type": "choice", "_value": values}
            ss.append(row)

        return ss
```

Next, we define a method build that constructs a search space of all cells according to the NNI specification:

```
@classmethod
def build(cls):
    return {
        "op_1": {
            "_type":  "choice",
            "_value": cls.operator_search_space('selector')
        },
        "op_2": {
            "_type":  "choice",
            "_value": cls.operator_search_space('transformer')
        },
        "op_3": {
            "_type":  "choice",
            "_value": cls.operator_search_space('transformer')
        },
        "op_4": {
            "_type":  "choice",
            "_value": cls.operator_search_space('transformer')
        },
        "op_5": {
            "_type":  "choice",
            "_value": cls.operator_search_space('classifier')
        }
    }
```

Even though the search space definition is quite large, we can print it out:

```
if __name__ == '__main__':
    search_space = SearchSpace.build()
    print(search_space)
```

We will use dynamic search space construction to launch the experiment in embedded mode, although this technique can also be applied to build a static JSON file.

Model

So, what do we have by now? We have an operator space and a search space. Let's now implement a model that converts the pipeline configuration into a real machine learning classifier. Listing 3-20 introduces `MlPipelineClassifier`.

Importing modules:

Listing 3-20. MlPipelineClassifier. ch3/ml_pipeline/model.py

```
from sklearn.pipeline import Pipeline
from ch3.ml_pipeline.operator import OperatorSpace
from ch3.ml_pipeline.utils import telescope_dataset

class MlPipelineClassifier:
```

Model receives pipeline configuration and converts it to the actual Scikit-learn pipeline:

```
    def __init__(self, pipe_config):

        ops = []
        for _, params in pipe_config.items():
            # operator name
            op_name = params.pop('_name')

            # skips 'none' operator
            if op_name == 'none':
                continue

            op = OperatorSpace.get_operator_by_name(op_name)
            ops.append((op.name, op.clz(**params)))

        self.pipe = Pipeline(ops)
```

Model training method:

```
    def train(self, X, y):
        self.pipe.fit(X, y)
```

Computing the accuracy:

```
    def score(self, X, y):
        return self.pipe.score(X, y)
```

Since the model is ready, let's try to initialize it using the sample pipeline parameter and apply it to the classification problem:

```
if __name__ == '__main__':

    pipe_config = {
        'op_1': {
            '_name':       'SelectPercentile',
            'percentile': 2
        },
        'op_2': {
            '_name': 'none'
        },
        'op_3': {
            '_name': 'Normalizer',
            'norm':  'l1'
        },
        'op_4': {
            '_name':            'PCA',
            'svd_solver':      'randomized',
            'iterated_power': 3
        },
        'op_5': {
            '_name':      'DecisionTreeClassifier',
            'criterion': "entropy",
            'max_depth': 8
        }
    }

    model = MlPipelineClassifier(pipe_config)
    X_train, y_train, X_test, y_test = telescope_dataset()
```

```
model.train(X_train, y_train)
score = model.score(X_test, y_test)
print(score)
```

The model demonstrates **64%** accuracy. This is definitely not the result we expect, so let's use the HPO techniques to construct a better-performance model.

Tuner

Now everything is ready to start the experiment. But I would like to focus on the search space that we use. The trial parameter is a sequence of operators, which may contain empty operators, that is, `S3(p3)` → none → `T1(p1)` → none → `C2(p2)`. But at the same time, the following parameter exists too: `S3(p3)` → none → none → `T1(p1)` → `C2(p2)`. They are different parameters in the search space but generate the same classifier model:

- `S3(p3)` → none → `T1(p1)` → none → `C2(p2)`
- `S3(p3)` → none → none → `T1(p1)` → `C2(p2)`
- `S3(p3)` → `T1(p1)` → `C2(p2)`.

Keep in mind that the same two identical operators with different parameters are not equal, that is, `SelectFwe(alpha=0)` is not equal to `SelectFwe(alpha=0.05)`. Let's customize Tuner by forbidding it to create parameters that will generate equivalent models concerning the parameters already tried, that is, if we already tried parameter P_1 = `SelectPercentile(percentile = 2)` → none → `Normalizer(norm='l1')` → none → `DecisionTreeClassifier(max_depth=8)`, then parameter P_2 = `SelectPercentile(percentile = 2)` → none → none → `Normalizer(norm='l1')` → `DecisionTreeClassifier(max_depth=8)` will not be passed to Experiment, because the model generated by P_2 equals to the model generated by P_1. Let's create `EvolutionShrinkTuner`, which inherits `EvolutionTuner` and tracks all executed pipelines forbidding passing the equal ones to the Experiment. We can see `EvolutionShrinkTuner` implementation in Listing 3-21.

Listing 3-21. EvolutionShrinkTuner. ch3/ml_pipeline/evolution_shrink_tuner.py

```
import json
from nni.algorithms.hpo.evolution_tuner import EvolutionTuner
class EvolutionShrinkTuner(EvolutionTuner):

    def __init__(self, optimize_mode = "maximize", population_size = 32):
        super().__init__(optimize_mode, population_size)
```

We define registry property that will track all created pipelines:

```
        self.registry = []
```

If the `super().generate_parameters` method of the parent `EvolutionTuner` object returns a parameter that has already been tried, then the `super().generate_parameters` method is called again until a unique parameter is generated. Because `EvolutionTuner` has random behavior, `super().generate_parameters` can be expected to return different parameters on subsequent calls.

```
    def generate_parameters(self, *args, **kwargs):
        params = super().generate_parameters(*args, **kwargs)

        # If not `params` are not valid generate new ones
        while not self.is_valid(params):
            params = super().generate_parameters(*args, **kwargs)

        return params
```

The following `is_valid` method converts the parameter to the canonical form by removing none operators and checks if it has already been tried: if it has, then returns `False`; if not, then saves it and returns `True`.

```
    def is_valid(self, params):
        # All step names
        step_names = [v['_name'] for _, v in params.items() if v['_name']
        != 'none']

        # No duplicates allowed
        if len(step_names) != len(set(step_names)):
            return False
```

```
    # `params` to canonical string
    canonical_form = 'X'
    for _, step_config in params.items():
        if step_config['_name'] == 'none':
            continue
        canonical_form += '--->' + json.dumps(step_config)

    # If `canonical_form` already tested
    if canonical_form in self.registry:
        return False

    self.registry.append(canonical_form)

    return True
```

This simple technique introduces the concept of equivalence between the elements of the search space and can significantly shrink the search space.

Experiment

And now, we are finally ready to launch an experiment to find the optimal functional pipeline for solving the AutoML problem. The trial script in Listing 3-22 initializes the model, prepares datasets, trains the model, tests it, and returns model accuracy to NNI Experiment.

(Full code is provided in the corresponding file: *ch3/ml_pipeline/trial.py*.)

Listing 3-22. Trial

```
def trial(hparams):
    #Initializing model
    model = MlPipelineClassifier(hparams)

    # Preparing dataset
    X_train, y_train, X_test, y_test = telescope_dataset()

    model.train(X_train, y_train)

    # Calculating `score` on test dataset
    score = model.score(X_test, y_test)
```

```
    # Send final score to NNI
    nni.report_final_result(score)
```

Listing 3-23 puts everything together and runs the experiment.

Import modules:

Listing 3-23. Optimal functional pipeline experiment. ch3/ml_pipeline/run_experiment.py

```
from pathlib import Path
from nni.experiment import Experiment, CustomAlgorithmConfig
from ch3.ml_pipeline.search_space import SearchSpace
```

Common Experiment configuration:

```
experiment = Experiment('local')
experiment.config.experiment_name = 'AutoML Pipeline'
experiment.config.trial_concurrency = 4
experiment.config.max_trial_number = 500
```

Generating search space:

```
experiment.config.search_space = SearchSpace.build()
```

Trial configuration:

```
experiment.config.trial_command = 'python3 trial.py'
experiment.config.trial_code_directory = Path(__file__).parent
```

Setting the custom EvolutionShrinkTuner:

```
experiment.config.tuner = CustomAlgorithmConfig()
experiment.config.tuner.code_directory = Path(__file__).parent
experiment.config.tuner.class_name = 'evolution_shrink_tuner.
EvolutionShrinkTuner'
experiment.config.tuner.class_args = {
    'optimize_mode':   'maximize',
    'population_size': 64
}
```

Launching Experiment:

```
http_port = 8080
experiment.start(http_port)

# Event Loop
while True:
    if experiment.get_status() == 'DONE':
        search_data = experiment.export_data()
        search_metrics = experiment.get_job_metrics()
        input("Experiment is finished. Press any key to exit...")
        break
```

Figure 3-41 shows top trials of the AutoML Experiment.

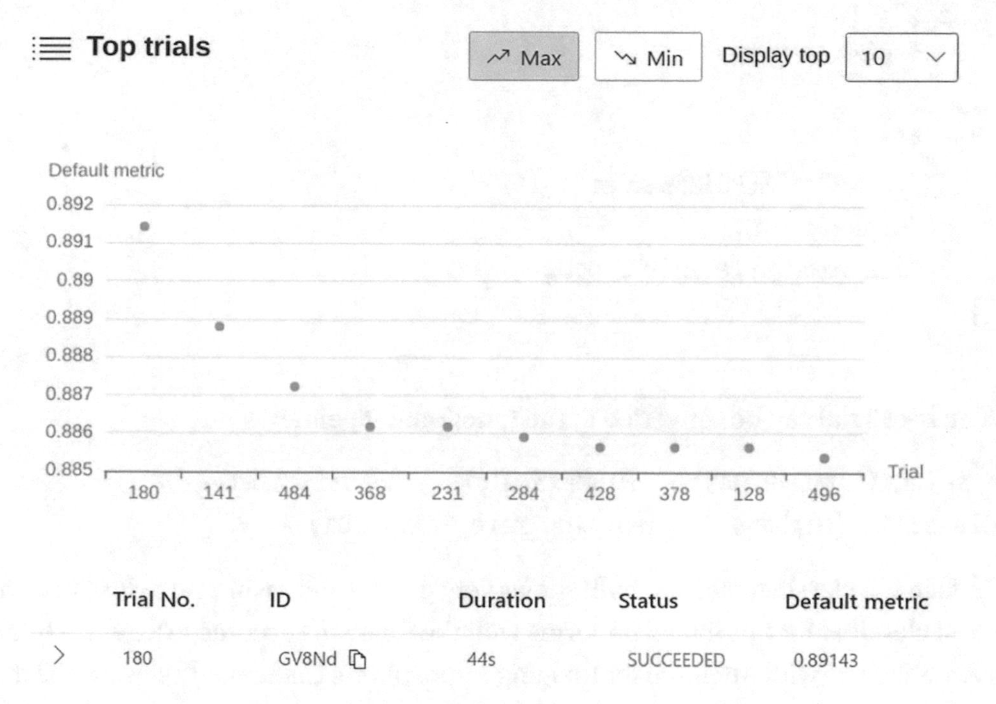

***Figure 3-41.** AutoML Experiment best Trials*

The best Trial demonstrates **0.89143** accuracy and has the following parameters:

```
{
    "op_1": {
        "_name": "SelectFwe",
        "alpha": 0.049
    },
    "op_2": {
        "_name": "MinMaxScaler"
    },
    "op_3": {
        "_name": "RobustScaler"
    },
    "op_4": {
        "_name": "none"
    },
    "op_5": {
        "_name": "MLPClassifier",
        "alpha": 0.01,
        "learning_rate_init": 0.01
    }
}
```

The best Trial can be converted to the functional pipeline:

```
X → Select(alpha=0.049) → MinMaxScaler → RobustScaler →
MLPClassifier(alpha=0.01, learning_rate_init=0.01) → Y
```

In fact, the classifier we have built shows very good results, which are close to optimal. The best classifier for this model performs with the following accuracy: `0.898` ("Multi-Task Architecture with Attention for Imaging Atmospheric Cherenkov Telescope Data Analysis," *www.scitepress.org/Papers/2021/102974/102974.pdf*). We have just built the custom AutoML toolkit based on NNI. This approach can also be applied to any functional pipeline optimization. Similarly, you can automatically design deep learning models with a sequential layer layout. Indeed, operator space can consist of deep learning layers, and the model can be a neural network based on the sequence of layers pipeline. Of course, we could dive deeper into applying this approach to Neural Architecture Search, but it has significant drawbacks, which we will discuss in the next section.

Limits of HPO Applying to Neural Architecture Search

The approach described in this section can be successfully applied to the search for neural network architectures but with a significant limitation. Using HPO techniques described in this section, we can only get an architecture with a sequential layer stack, that is, an architecture in which each layer connects to another one sequentially. This is a significant limitation since many modern neural network architectures use multiple connections between their layers, as shown in Figure 3-42.

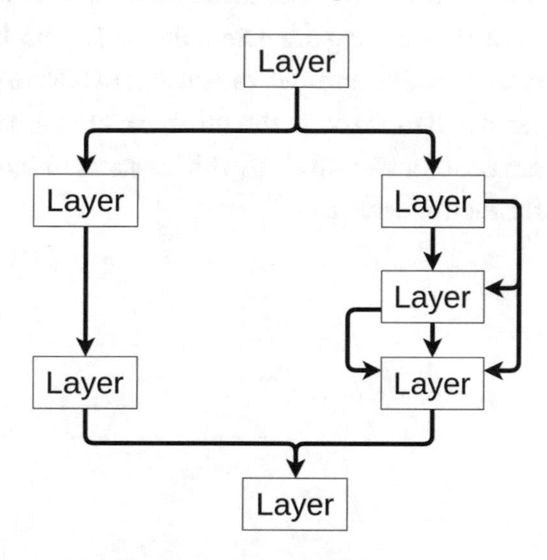

***Figure 3-42.** Multi-connected neural network*

We need some different and special techniques to search for efficient neural network architectures. And we will begin to explore them in the next chapter.

Hyperparameters for Hyperparameter Optimization

When we talked about automated deep learning, we mentioned that AutoDL solves the problem of finding optimal architectures and hyperparameters. But each AutoDL technique has many parameters on its own! For example, we need to define the search space, a maximum number of trials, find a suitable Tuner, define Tuner parameters, etc. It turns out that the problem that solves the hyperparameter selection problem itself has even more hyperparameters! What is the point of using the HPO approach and other AutoDL methods then? This is a fair question. The answer is that even a poorly configured HPO Experiment produces a significantly better model than a model with a poorly configured architecture or hyperparameters. We can never be sure in advance what the best settings for an AutoDL method are. However, we can be confident that the AutoDL method will surely produce a model close to the optimal one and significantly better than the one we would build manually.

Summary

This chapter has taken a deep dive into Tuner internals and various black-box function optimization algorithms. Understanding the principles of Tuners' behavior and their practical application can remarkably improve the design of NNI experiments and HPO results. In the next chapter, we'll move on to the most exciting and interesting part of our book: Neural Architecture Search. We will study the latest techniques to find the optimal design of neural networks for a specific task.

CHAPTER 4

Multi-trial Neural Architecture Search

And now we come to the most exciting part of this book. As we noted at the end of the last chapter, HPO methods are pretty limited for automating the search for the optimal deep learning models, but **Neural Architecture Search** (**NAS**) dispels these limits. This chapter focuses on NAS, one of the most promising areas of automated deep learning. Automatic Neural Architecture Search is increasingly important in finding appropriate deep learning models. Recent researches have proven the NAS effectiveness and found some models that could beat manually tuned ones. NAS is a fairly young discipline in machine learning. It took shape as a separate discipline in 2018. Since then, it has made a significant breakthrough in automating neural network architecture construction that solves a specific problem. The most manual design of neural networks can be replaced by automated architecture search soon, so this area is very up and coming for all data scientists. NAS produced many top computer vision architectures. Architectures like NASNet, EfficientNet, and MobileNet are the result of automated Neural Architecture Search.

There are two types of NAS: ***Multi-trial*** and ***One-shot***. In Multi-trial NAS, a model evaluator evaluates each sampled model's performance, and an Exploration Strategy samples models from defined Model Space, while One-shot NAS tries to find optimal neural architecture training and exploring one Supernet derived from the Model Space. This chapter is dedicated to Multi-trial NAS.

This chapter is divided into two parts: ***Neural Architecture Search Using Retiarii (PyTorch)*** and ***Classic Neural Architecture Search (TensorFlow)***. ***Retiarii*** is a deep learning framework that supports the exploratory training on a neural network Model Space developed by NNI experts. Retiarii is an advantageous approach that allows structuring and planning the NAS. Unfortunately, the NNI `2.7` version (which is used in this book) only implements the Retiarii approach for the PyTorch framework. And it would be unfair not to pay attention to the TensorFlow framework in this chapter, so

I. Gridin, *Automated Deep Learning Using Neural Network Intelligence*,
https://doi.org/10.1007/978-1-4842-8149-9_4

the classic methods of NAS using NNI are considered in the *Classic Neural Architecture Search (TensorFlow)* part. In any case, NNI supports TensorFlow for One-shot NAS, which we will explore in the next chapter. Therefore, TensorFlow users will be able to take full advantage of NAS approaches.

Neural Architecture As Data Flow Graph

We will begin this chapter by defining how the NAS perceives the neural network architecture. Neural architecture is considered as a Data Flow Graph (DFG). DFG is a collection of nodes and connections between them. DFG displays the data transfer from one node to another. Each node has its own type and parameters. Figure 4-1 demonstrates an example of DFG.

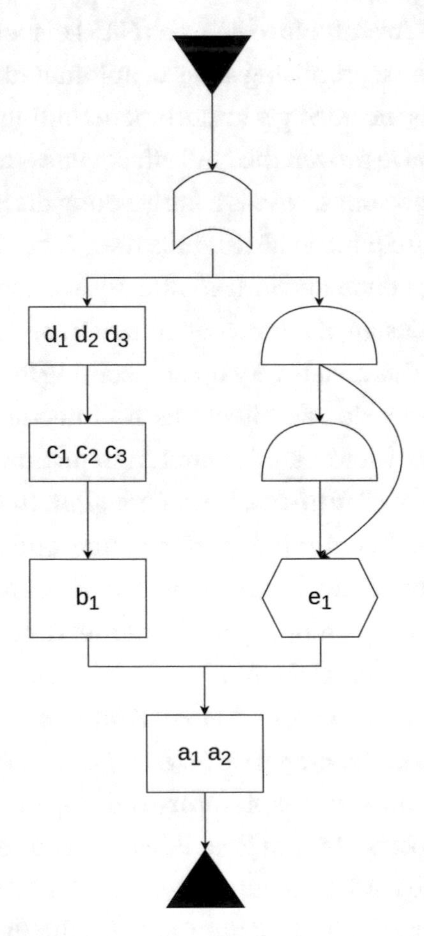

Figure 4-1. *Data Flow Graph*

In Figure 4-1, we see the DFG, which contains different types of nodes with parameters. The architecture of each neural network can be represented as a Data Flow Graph. Indeed, in Figure 4-1, we can replace the rectangle node with a convolution layer with parameters: `padding`, `stride`, `filter_size`, etc. Neural Architecture Search Explorers have no idea about the nature of each of the DFG nodes. The main task is to construct a DFG from various deep learning layers, which forms the architecture of a neural network and solves a specific problem in the best way.

Neural Architecture Search Using Retiarii (PyTorch)

Retiarii is the framework developed by NNI experts, and it is the first framework that supports deep learning exploratory training. Exploratory training implies that different deep neural networks (DNNs) training results are shared. This approach compares training of different models, performs optimization, and stops unpromising models with poor intermediate results. Also, Retiarii provides a new interface to specify a deep learning Model Space for exploration and an interface to describe the Exploration Strategy that decides the order to instantiate and train models in, prioritize model training, and stop the training of certain models. Retiarii identifies the correlations between the instantiated models and develops a set of cross-model optimizations to improve the overall exploratory training process. You can read more information about the Retiarii framework in the following article: `www.usenix.org/system/files/osdi20-zhang_quanlu.pdf`.

NNI version `2.7` has a PyTorch-only implementation of the Retiarii framework. In the next NNI releases, the TensorFlow implementation should be added. Please refer to the official documentation to check the actual state: `https://nni.readthedocs.io/en/v2.7/nas/overview.html`.

Introduction to NAS Using Retiarii

The best way to dive into NAS using Retiarii is to examine a simple example of finding the optimal DFG for a particular task. We will not go deep into the design of neural networks but will focus on a simple arithmetic problem. Suppose we have some chain of operators *F*, which performs the following actions:

```
x = 1
x ← x × 2
x ← x × 4
return x
```

We can enrich the chain *F* with the following statements:

```
x = 1
x ← sigmoid(x) or x ← tanh(x) or x ← relu(x)
x ← x × 2
x ← x + 0 or x ← x + 1 or x ← x + 2
x ← x × 4
x ← sigmoid(x) or x ← tanh(x) or x ← relu(x)
return x
```

The Model Space of this problem can be depicted as shown in Figure 4-2. We need to find a DFG that maximizes output.

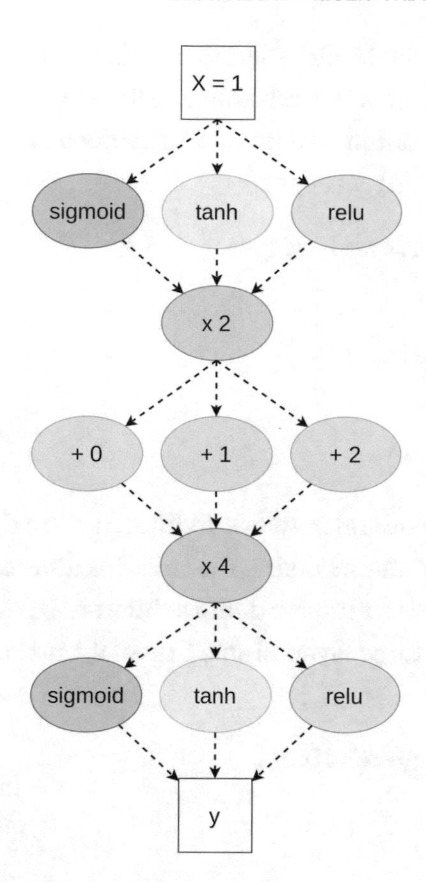

Figure 4-2. *Data Flow Graph Model Space*

Of course, this is a trivial task, but it is well suited as an introductory example of using the Retiarii framework in NNI. Let's move on to solving this problem. Listing 4-1 defines the Model Space.

We import common modules:

Listing 4-1. Model Space. ch4/retiarii/intro/dummy_model.py

```
import os
import torch
import nni
```

Retiarii uses import `nni.retiarii.nn.pytorch as nn` for PyTorch. It is very important to use layers from this package implementing NAS for PyTorch.

```
import nni.retiarii.nn.pytorch as nn
```

First, we need to define the Model Space for Exploration. Model Space in the NAS context can be considered as search space in the HPO context. Model Space is represented by a class that contains all possible architectures to try. Each such class must be annotated with `@model_wrapper`.

```
from nni.retiarii import model_wrapper

@model_wrapper
class DummyModel(nn.Module):

    def __init__(self):
        super().__init__()
```

Different variants of the neural network architecture are determined by special methods called *Mutators*. Mutators define the rules for different variations or mutations of the model. The `LayerChoice` mutator defines different layer choices. In the following, `LayerChoice` selects one of three layers: `Tanh`, `Sigmoid`, and `ReLU`:

```
        # operator 1
        self.op1 = nn.LayerChoice([
            nn.Tanh(),
            nn.Sigmoid(),
            nn.ReLU()
        ])
```

Another type of mutation is `ValueChoice`. This mutator selects one of the values from the list:

```
        # addition
        self.add = nn.ValueChoice([0, 1, 2])
```

Next, we define the `LayerChoice` mutator again:

```
        # operator 2
        self.op2 = nn.LayerChoice([
            nn.Tanh(),
            nn.Sigmoid(),
            nn.ReLU()
        ])
```

Finally, we chain all operators together, as shown in Figure 4-2:

```
def forward(self, x):
    x = self.op1(x)
    x = x * 2
    x += self.add
    x = x * 4
    x = self.op2(x)
    return x
```

After, we define the evaluate method, which returns the model result:

```
def evaluate(model_cls):
```

Evaluating the model:

```
model = model_cls()
x = torch.Tensor([1])
y = model(x)
```

This code is used for model architecture visualization. We will get back to this technique later.

```
onnx_dir = os.path.abspath(os.environ.get('NNI_OUTPUT_DIR', '.'))
os.makedirs(onnx_dir, exist_ok = True)
torch.onnx.export(model, x, onnx_dir + '/model.onnx')
```

Returning the result:

```
nni.report_final_result(y.item())
```

Once we have defined the Model Space and its instance Evaluation, we can move on to launching the experiment with the code shown in Listing 4-2.

Importing necessary modules:

Listing 4-2. Retiarii Experiment. ch4/retiarii/intro/run_experiment.py

```
from time import sleep
from nni.retiarii.evaluator import FunctionalEvaluator
from nni.retiarii.experiment.pytorch import RetiariiExperiment, RetiariiExeConfig
import nni.retiarii.strategy as strategy
from ch4.retiarii.intro.dummy_model import DummyModel, evaluate
```

There are seven main steps to launch the Retiarii Experiment:

1. We set the Model Space:

   ```
   model_space = DummyModel()
   ```

2. We define the Evaluator, which evaluates model instance:

   ```
   evaluator = FunctionalEvaluator(evaluate)
   ```

3. Next, we chose a Search Strategy that explores the Model Space:

   ```
   search_strategy = strategy.Random(dedup = True)
   ```

4. We initialize the Retiarii Experiment with defined Model Space, Evaluator, and Search Strategy:

   ```
   exp = RetiariiExperiment(model_space, evaluator, [], search_
   strategy)
   ```

5. Setting Experiment configuration:

   ```
   exp_config = RetiariiExeConfig('local')
   exp_config.experiment_name = 'dummy_search'
   exp_config.trial_concurrency = 1
   exp_config.max_trial_number = 100
   exp_config.training_service.use_active_gpu = False
   export_formatter = 'dict'
   ```

6. Launching Experiment:

   ```
   exp.run(exp_config, 8080)
   ```

7. Returning best results. We can analyze the Experiment's results in WebUI before exit:

   ```
   while True:
       sleep(1)
       input("Experiment is finished. Press any key to exit...")
       print('Final model:')
   ```

```
for model_code in exp.export_top_models(formatter =
export_formatter):
    print(model_code)
break
```

After the experiment is completed, we can analyze the Trial jobs panel examining the best results.

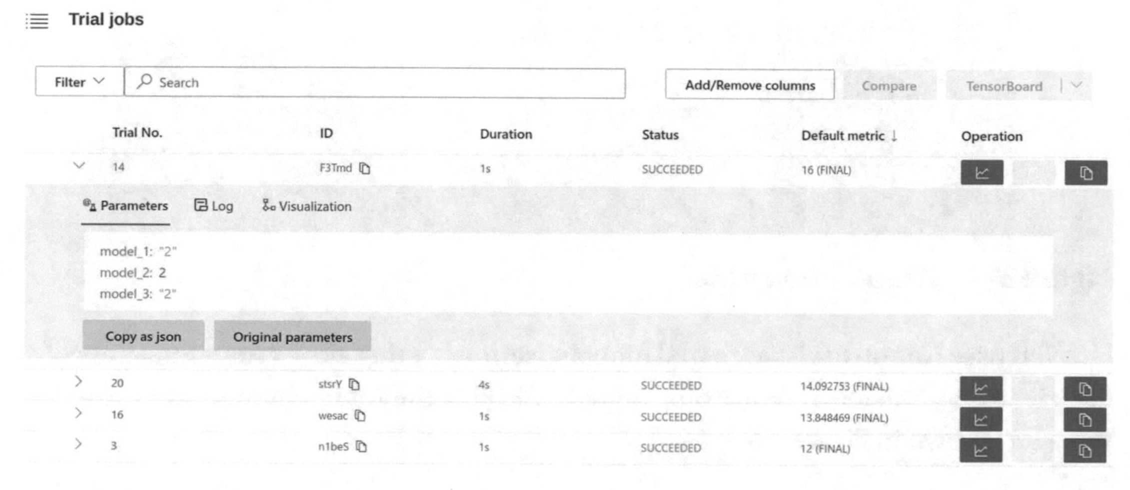

***Figure 4-3.** Trial jobs panel*

As shown in Figure 4-3, the best model returns 16. And it has the following set of parameters:

```
{
    "model_1": "2",
    "model_2": 2,
    "model_3": "2"
}
```

The preceding parameters are not self-describing, so we can use the visualization function to render DFG render as shown in Figure 4-4.

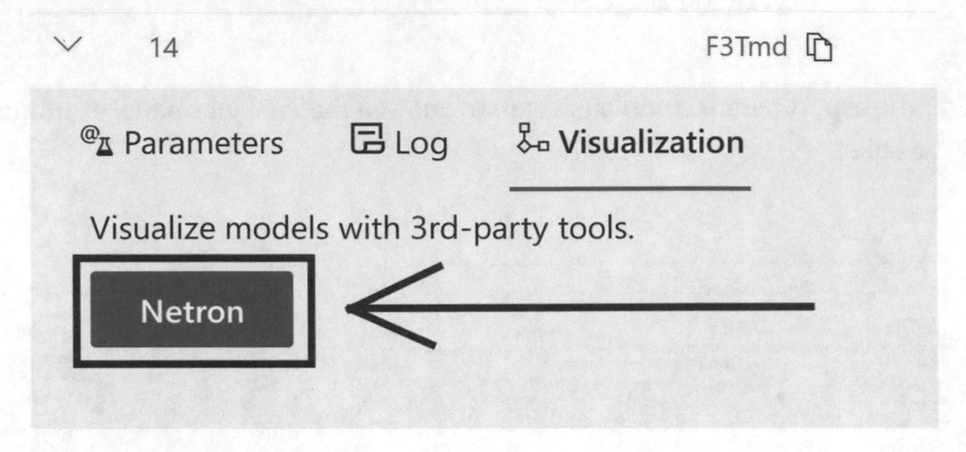

Figure 4-4. *Visualization panel*

NNI uses Netron to visualize trial models. Netron is a tiny viewer for neural networks, deep learning, and machine learning models. Clicking the `Netron` button, you'll see the screen as shown in Figure 4-5.

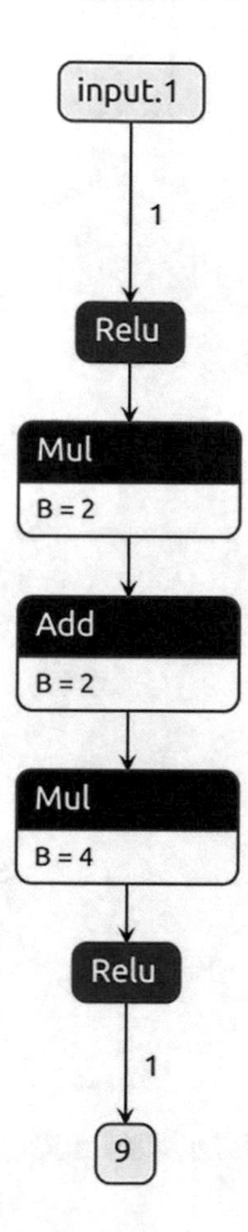

Figure 4-5. *Netron model visualization*

Now we can declare that we have found a solution to the problem of finding a model that maximizes the value of the chain of operators *F*. The Data Flow Graph of this model is shown in Figure 4-6.

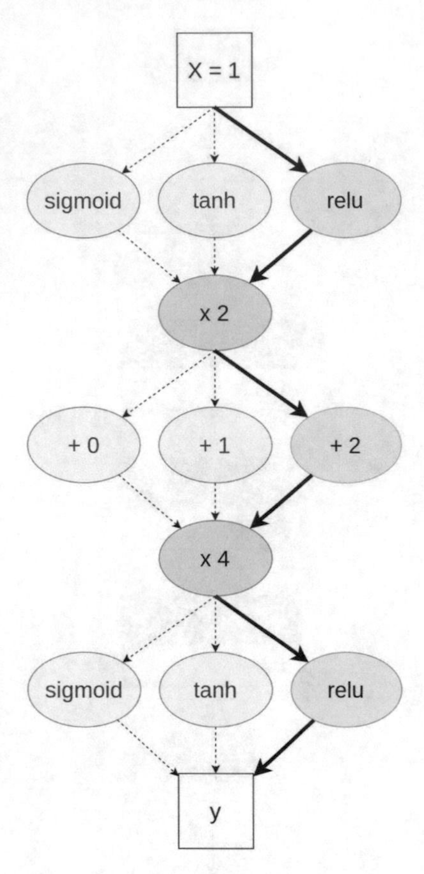

Figure 4-6. *Best Data Flow Graph in Model Space*

The purpose of this example was to demonstrate that the primary goal of a NAS approach is to find a DFG that maximizes (or minimizes) a black-box function. In the same way HPO methods are searching for parameters that maximize (or minimize) the black-box function. After introducing basic NAS techniques, we can dive into more details.

Retiarii Framework

Retiarii framework is designed to separate the main logical entities of Neural Architecture Search. This makes the NAS procedure clear and elegant. Using the Retiarii framework, the researcher can only focus on particular aspects of the investigation. The main components of the Retiarii framework are

- Base Model
- Mutator
- Model Space
- Evaluator
- Exploration Strategy

Base Model is the primary skeleton of a neural network. Base Model is actually a simple deep learning model that solves some problem. Often the Base Model does not show good performance. But it has some primary neural architecture and training algorithm.

Mutator is a possible change that a Base Model can be subjected to. Mutator defines the transformation of the Base Model architecture into another one. Usually, many mutators are applied to Base Model.

Model Space is the set of all possible Base Model mutations. Each mutator generates several variants of neural network architectures. Applying all mutators to the Base Model defines the Model Space.

Evaluator measures the performance of a sample from the Model Space. This is a typical algorithm for training and testing a neural network.

Exploration Strategy defines the Model Space exploration algorithm. The main objective of the Exploration Strategy is to find the best model in the least number of trials.

All these concepts are pretty familiar to us after studying Hyperparameter Optimization. Table 4-1 contains the main logical entities from NAS and HPO. As you can see, they mean almost the same thing.

Table 4-1. *NAS and HPO logical entities*

NAS	HPO
Base Model + Mutator	Search space type
Model Space	Search space
Evaluator	Trial
Exploration Strategy	Tuner

Figure 4-7 illustrates the relationship of various components in the Retiarii framework.

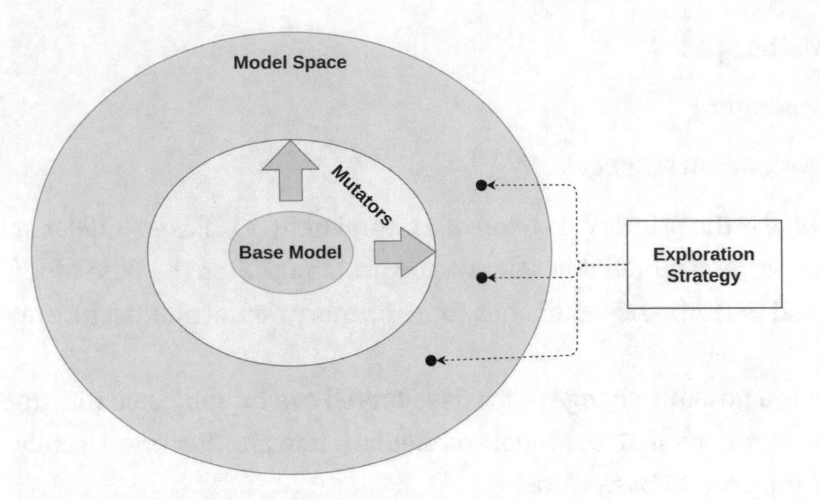

Figure 4-7. *Retiarii framework*

Let's go ahead and look at each of these components in detail.

Base Model

The Base Model is a starting point from which all possible architecture modifications will be made. For example, the Base Model for NAS of the MNIST problem can be presented as it is shown in Listing 4-3.

Importing PyTorch modules:

Listing 4-3. Base Model. ch4/retiarii/common/base_model.py

```
import torch
import torch.nn.functional as F
```

You must use the `nni.retiarii.nn.pytorch as nn` module to declare layers in a deep learning model:

```
import nni.retiarii.nn.pytorch as nn
```

Base Model must be annotated with nni.retiarii.model_wrapper():

```
from nni.retiarii import model_wrapper
@model_wrapper
class Net(nn.Module):
```

Next, we have the classic LeNet model design for the digit recognition problem:

```
    def __init__(self):
        super().__init__()
        self.conv1 = nn.Conv2d(1, 32, 3, 1)
        self.conv2 = nn.Conv2d(32, 64, 3, 1)
        self.dropout1 = nn.Dropout(0.25)
        self.dropout2 = nn.Dropout(0.5)
        self.fc1 = nn.Linear(9216, 128)
        self.fc2 = nn.Linear(128, 10)

    def forward(self, x):
        x = F.relu(self.conv1(x))
        x = F.max_pool2d(self.conv2(x), 2)
        x = torch.flatten(self.dropout1(x), 1)
        x = self.fc2(self.dropout2(F.relu(self.fc1(x))))
        output = F.log_softmax(x, dim = 1)
        return output
```

Frequent mistakes in Base Model design are:

- Missing @model_wrapper annotation on Base Model
- Using import torch.nn as nn instead of nni.retiarii.nn.pytorch as nn declaring layers

Mutators

Base Model is a single model. To create Model Space, we have to add Mutators to the Base Model. Each mutator provides a way to change the Base Model. All possible mutations applied to the Base Model form the Model Space. NNI provides the following mutation operations: LayerChoice, ValueChoice, InputChoice, and Repeat.

LayerChoice

LayerChoice mutator forms the candidate layers for a layer placeholder. One of these layers is tried in the exploration process. LayerChoice mutator is applied to the Base Model the following way:

```
# import part
import nni.retiarii.nn.pytorch as nn

# model design
self.activation = nn.LayerChoice([
    nn.ReLU(),
    nn.Sigmoid(),
    nn.Identity
])

# forward
x = self.activation(x)
```

LayerChoice adds layer variations to the Base Model as shown in Figure 4-8.

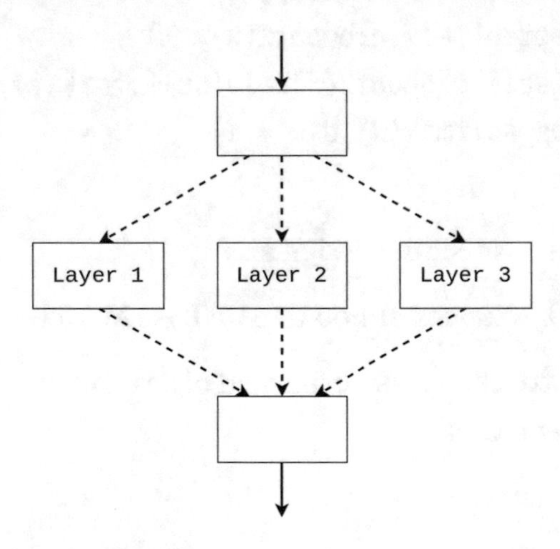

Figure 4-8. *LayerChoice Mutator*

LayerChoice is the most straightforward way to mutate the Base Model.

ValueChoice

ValueChoice forms the list of single values to be tried as layer hyperparameters. ValueChoice can be used as layer hyperparameters only. It cannot be used as an arbitrary hyperparameter like batch_size or learning_rate in the evaluation process. ValueChoice mutator is applied to the Base Model the following way:

```
# import part
import nni.retiarii.nn.pytorch as nn

# model design
self.drop = nn.Dropout(nn.ValueChoice([0.25, 0.5, 0.75]))

# forward
x = self.drop(x)
```

ValueChoice can be considered as a layer hyperparameter in the HPO context.

InputChoice

InputChoice tries different connections. It takes several tensors and chooses n_chosen tensors from them. InputChoice mutator is applied to the Base Model the following way:

```
# import part
import nni.retiarii.nn.pytorch as nn

# model design
self.switch = nn.InputChoice(n_candidates = 2, n_chosen = 1)

# forward

# branch one
a = self.op_a1(x)
a = self.op_a2(a)

# branch two
b = self.op_b1(x)
b = self.op_b2(b)

# choosing connection
x = self.switch([a, b])
```

InputChoice is designed to find the best data flow branches in neural network architecture. Figure 4-9 illustrates this concept.

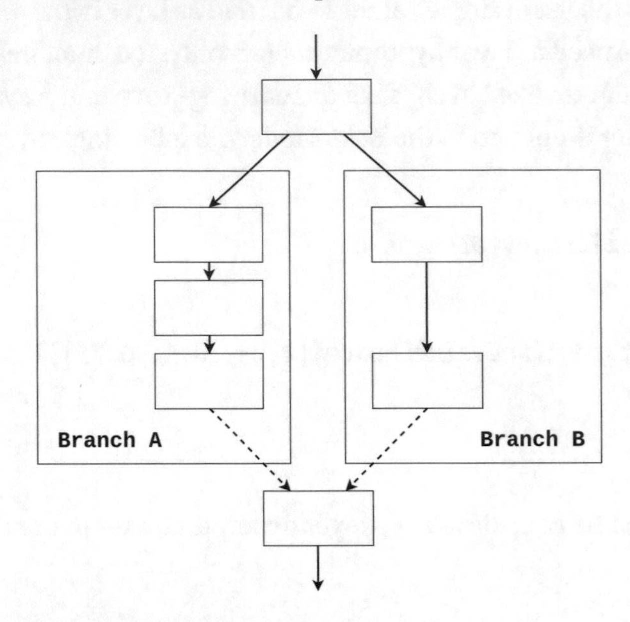

Figure 4-9. *InputChoice Mutator*

If InputChoice picks more than one candidate tensors (i.e., n_chosen > 1), then the reduction strategy is applied: sum, mean, concat. This is a very useful technique that allows to extract and merge several connections at the same time. InputChoice for multiple candidates with reduction can be applied the following way:

```
# import part
import nni.retiarii.nn.pytorch as nn

# model design
self.mix = nn.InputChoice(n_candidates = 3, n_chosen = 2, reduction = 'sum')

# forward

# branch one
a = self.op_a1(x)
a = self.op_a2(a)
```

```
# branch two
b = self.op_b1(x)
b = self.op_b2(b)

# branch three
c = self.op_b1(x)
c = self.op_b2(c)

# choosing connection
x = self.mix([a, b, c])
```

The preceding code generates the search space shown in Figure 4-10.

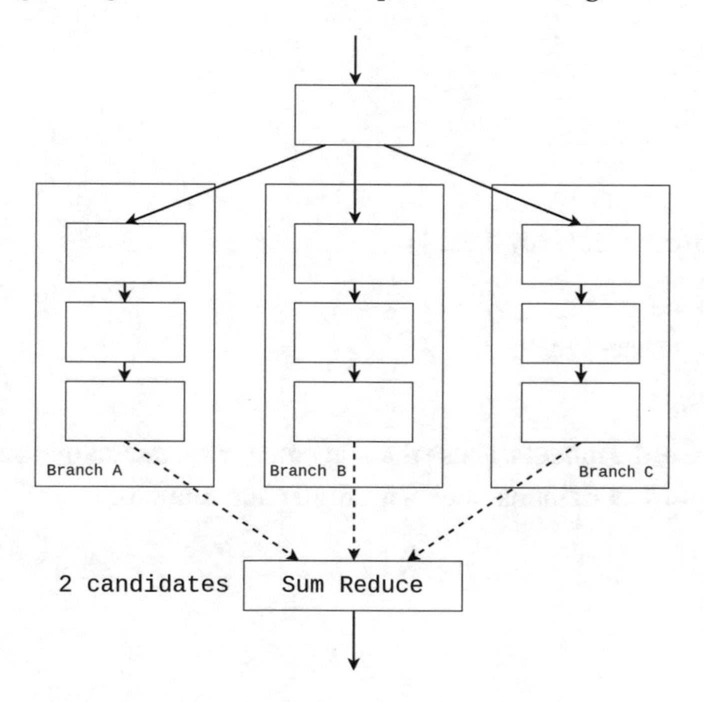

Figure 4-10. *InputChoice Mutator for multiple candidates*

`InputChoice` Mutator for multiple candidates can choose the same tensor several times. It happens when other connections do not bring any helpful information to neural network performance.

Also, InputChoice allows adding skip connection technique to neural architecture. Skip connection is one of the core techniques in modern neural network design. It was first introduced in 2015 in "Deep Residual Learning for Image Recognition" (https://arxiv.org/pdf/1512.03385.pdf). Skip connection can be implemented using the following pattern:

```
# import part
import nni.retiarii.nn.pytorch as nn

# model design
self.skip_connect = nn.InputChoice(n_candidates = 2, n_chosen = 1)

# forward
x0 = x.clone()

# connection
x1 = self.op(x)

x0 = self.skip_connect([x0, None])

if x0 is not None:
    # skipping connection
    x1 += x0
```

In the first case, the model will use the skip connection technique, but not in the second one. Figure 4-11 demonstrates skip connection mutation.

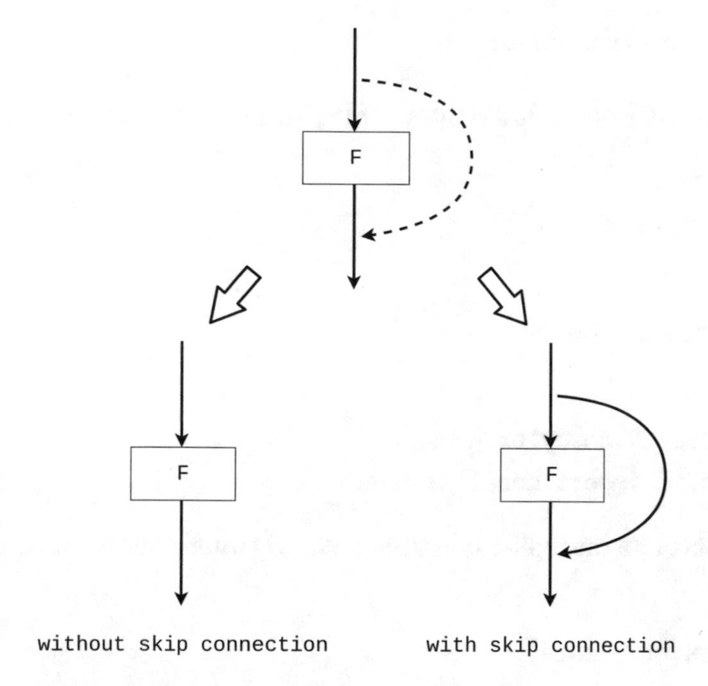

Figure 4-11. *InputChoice. Skip connection mutation*

Let's examine the application of the `InputChoice` mutator using an arithmetic DFG as an example. Say we have $x = 1$ and three operator pipelines which consist of multiplication operators:

```
1: x → (x × 2)
2: x → (x × 2) → (x × 3)
3: x → (x × 2) → (x × 3) → (x × 4)
```

You need to select two pipelines whose sum is maximum. You can select the same pipeline twice. This is a fairly simple task. It is intuitively clear that the sum of the last pipeline will give the maximum value. Let's run the third pipeline: $1 \rightarrow 1 \times 2 \rightarrow 2 \times 3 \rightarrow 6 \times 4 \rightarrow 24$; thus, the maximum value we can obtain is 48.

Now let's obtain the same result using NNI and `InputChoice` Mutator using Listing 4-4.

Importing common modules:

Listing 4-4. InputChoice Model Space. ch4/retiarii/common/input_choice/model_space.py

```
import os
import torch
```

Importing nni modules:

```
import nni
import nni.retiarii.nn.pytorch as nn
from nni.retiarii import model_wrapper
```

ProdBlock acts as a multiplier operator. It simply multiplies the tensor by some value:

```
class ProdBlock(nn.Module):

    def __init__(self, multiplier = 0):
        super().__init__()
        self.multiplier = multiplier

    def forward(self, x):
        x = x * self.multiplier
        return x
```

Next, we define the Model Space:

```
@model_wrapper
class InputChoiceModelSpace(nn.Module):

    def __init__(self):
        super().__init__()
```

We declare multiplier operators

```
        self.x2 = ProdBlock(2)
        self.x3 = ProdBlock(3)
        self.x4 = ProdBlock(4)
```

and InputChoice mutator that will select the best pair of pipelines:

```
        self.mix = nn.InputChoice(
            n_candidates = 3,
            n_chosen = 2,
            reduction = 'sum'
        )
```

forward action executes three different pipelines, and InputChoice mutator tries only two of them:

```
    def forward(self, x):
```

First pipeline: x → (x × 2)

```
        # Branch A
        a = self.x2(x)
```

Second pipeline: 2: x → (x × 2) → (x × 3)

```
        # Branch B
        b = self.x2(x)
        b = self.x3(b)
```

Third pipeline: x → (x × 2) → (x × 3) → (x × 4)

```
        # Branch C
        c = self.x2(x)
        c = self.x3(c)
        c = self.x4(c)

        return self.mix([a, b, c])
```

Here is the evaluation function:

```
def evaluate(model_cls):
    model = model_cls()
    x = 1
    out = model(x)
```

```
    # visualizing
    onnx_dir = os.path.abspath(os.environ.get('NNI_OUTPUT_DIR', '.'))
    os.makedirs(onnx_dir, exist_ok = True)
    torch.onnx.export(model, x, onnx_dir + '/model.onnx')

    nni.report_final_result(out)
```

Now we can run the experiment using this script:

```
$ python3 ch4/retiarii/common/input_choice/run_experiment.py
```

You can analyze the results on the WebUI detail page: `http://127.0.0.1:8080/detail`.

Trial No.	ID	Duration	Status	Default metric ↓
8	lRN6H	1s	SUCCEEDED	48 (FINAL)
2	yw3Zi	1s	SUCCEEDED	30 (FINAL)
6	rHRMh	1s	SUCCEEDED	30 (FINAL)
5	IJUoh	1s	SUCCEEDED	26 (FINAL)
7	MXzTz	1s	SUCCEEDED	26 (FINAL)
0	jbZhl	1s	SUCCEEDED	12 (FINAL)
1	NKAbE	1s	SUCCEEDED	8 (FINAL)
3	KoLyt	1s	SUCCEEDED	8 (FINAL)
4	Ft1gl	1s	SUCCEEDED	4 (FINAL)

Figure 4-12. *InputChoice Experiment results*

In Figure 4-12, we see $3^2 = 9$ trials. The best trial shows 48 and has the following parameters: `{ "model_1_0": 2, "model_1_1": 2 }`, which means that the best result is achieved with the last pipelines.

```
return self.mix(
    [
        a, # <- 0
        b, # <- 1
        c  # <- 2
    ])
```

Such simple examples help to understand better how mutators act before proceeding to a real NAS.

Repeat

Repeat mutator repeats some action a certain number of times. In the NAS context, the Repeat mutator tries to determine how often to iterate the same neural network block. For example, the ResNet neural network architecture implies a stack of Residual Blocks. But the optimal number of Residual Blocks may depend on the specific task. Figure 4-13 shows part of the ResNet architecture.

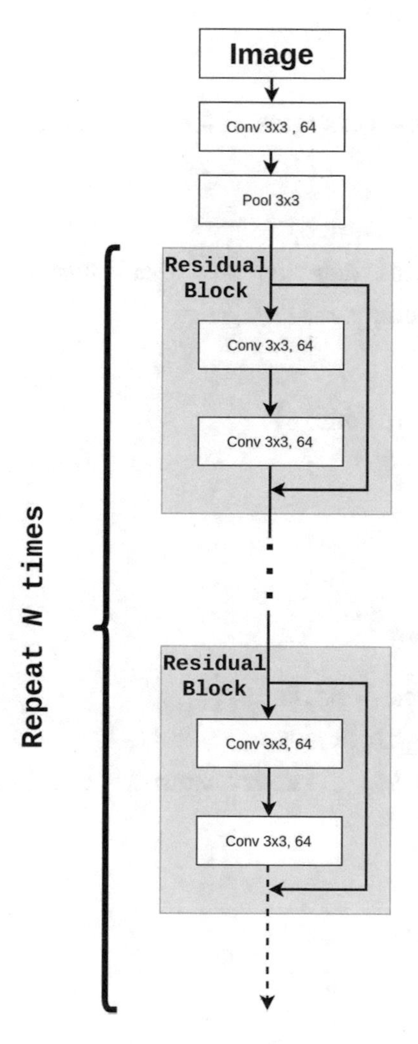

***Figure 4-13.** ResNet architecture*

Repeat mutator accepts a function that generates a block by its sequence number. Here is the pattern how Repeat mutator can be applied:

```
import nni.retiarii.nn.pytorch as nn
from nni.retiarii import model_wrapper
```

We define a custom model block:

```
class SomeBlock(nn.Module):
    ...
```

Set a builder function that generates a block concerning its ordinal number in the stack:

```
def create_some_block(block_num):
    # some logic here that depends on 'block_num'
    return SomeBlock(block_num)

@model_wrapper
class RepeatModelSpace(nn.Module):

    def __init__(self):
        super().__init__()
        ...
```

Defining Repeat mutator:

```
        self.repeat_block = nn.Repeat(
            create_some_block,
            depth = (1, 5)  # repeat from 1 to 5 times
        )
        ...

    def forward(self, x):
        ...
```

Evaluating Repeat mutator:

```
        x = self.repeat_block(x)
        ...
```

Therefore, it would be handy to have a tool that iteratively constructs block stacks of various lengths. And that's what the `Repeat` Mutator does. Let's examine the implementation of the `Repeat` mutator for a synthetic arithmetic task, as we did for the `InputChoice` mutator. Say we have a sequence of blocks that add values to a tensor. And we need to find the optimal length of this sequence. Listing 4-5 describes the Model Space for this problem.

`AddBlock` acts as an addition operator. It simply adds some value to the input tensor.

Listing 4-5. Repeat Mutator Model Space. ch4/retiarii/common/repeat/model_space.py

```
class AddBlock(nn.Module):

    def __init__(self, add = 0):
        super().__init__()
        self.add = add

    def forward(self, x):
        x = x + self.add
        return x
```

Builder function that creates `AddBlock` by its ordinal number:

```
    @classmethod
    def create(cls, block_num):
        return AddBlock(block_num)
```

Next, we define the Model Space:

```
@model_wrapper
class RepeatModelSpace(nn.Module):

    def __init__(self):
        super().__init__()
```

Repeat mutator generates block sequences of different lengths (from 1 to 5):

```
        self.repeat = nn.Repeat(
            AddBlock.create,
            depth = (1, 5)
        )

    def forward(self, x):
        return self.repeat(x)
```

You can examine the experiment by running

```
$ python3 ch4/retiarii/common/repeat/run_experiment.py
```

In fact, the primitive mutators LayerChoice, ValueChoice, InputChoice, and Repeat allow you to construct a space of any complexity. These mutators can be compared to programming language directives:

- *set*: LayerChoice, ValueChoice
- *if*: InputChoice
- *loop*: Repeat

Later in this chapter, we will examine a Model Space construction for a real NAS task using these mutators.

Labeling

All the mutator APIs have an optional argument label. Mutators with the same label will share the same value. A typical example is

```
self.net = nn.Sequential(
    nn.Linear(10, nn.ValueChoice([32, 64, 128], label='hidden_dim'),
    nn.Linear(nn.ValueChoice([32, 64, 128], label='hidden_dim'), 3)
)
```

which is the same as

```
hidden_dim = nn.ValueChoice([32, 64, 128], label='hidden_dim')
self.net = nn.Sequential(
    nn.Linear(10, hidden_dim,
    nn.Linear(hidden_dim, 3)
)
```

Example

Listing 4-6 demonstrates a trivial example of the Model Space applied to the image classification network.

Listing 4-6. Model Space. ch4/retiarii/common/model_space.py

```
@model_wrapper
class Net(nn.Module):

    def __init__(self):
        super().__init__()
        self.conv1 = nn.Conv2d(1, 32, 3, 1)
```

Applying LayerChoice mutator:

```
        self.conv2 = nn.LayerChoice([
            nn.Conv2d(32, 64, 3, 1),
            nn.Identity
        ], label = 'conv_layer')
```

Applying ValueChoice mutator as Dropout layer hyperparameter value:

```
        self.dropout1 = nn.Dropout(
            nn.ValueChoice([0.25, 0.5, 0.75]),
            label = 'dropout'
        )

        self.dropout2 = nn.Dropout(0.5)
```

Applying ValueChoice mutator as Linear hyperparameter value:

```
        feature = nn.ValueChoice(
            [64, 128, 256],
            label = 'hidden_size'
        )
        self.fc1 = nn.Linear(9216, feature)
        self.fc2 = nn.Linear(feature, 10)

    def forward(self, x):
        x = F.relu(self.conv1(x))
        x = F.max_pool2d(self.conv2(x), 2)
        x = torch.flatten(self.dropout1(x), 1)
        x = self.fc2(self.dropout2(F.relu(self.fc1(x))))
        output = F.log_softmax(x, dim = 1)
        return output
```

Evaluators

Retiarii Evaluator is a function that accepts a model class, initiates a model, trains it, tests it, and returns a result to Experiment. Evaluator can be implemented using the following pattern:

```
def evaluate(model_cls):
    # Initiate model
    model = model_cls()

    # Saving model graph for Visualization
    onnx_dir = os.path.abspath(os.environ.get('NNI_OUTPUT_DIR', '.'))
    os.makedirs(onnx_dir, exist_ok = True)
    torch.onnx.export(model, input_x, onnx_dir + '/model.onnx')

    # Model Training
    # ... passing intermediate results
    # ... nni.report_intermediate_result()

    # Model Testing
    # nni.report_final_result(out)
```

Retiarii Evaluator is pretty close to the Trial approach we used in HPO in previous chapters.

Exploration Strategies

NNI provides the following exploration strategies for Multi-shot NAS: *Random Strategy*, *Grid Search*, *Regularized Evolution*, *TPE Strategy*, and *RL Strategy*. We are already familiar with some of them because they implement the same approach as corresponding HPO Tuners. But anyway, let's briefly study each of them.

Random Strategy

Random Strategy (`nni.retiarii.strategy.Random`) randomly samples new models from the Model Space. It is a simple but still effective technique to Explore Model Space. Random Search is a good first-time Exploration Strategy, and it can give you good clues when you have no idea about the dataset you are dealing with and suitable architecture designs. Usually, Random Search is used first, and after the Model Space is refined, a more intelligent Exploration Strategy is applied.

Grid Search

Grid Search Strategy (`nni.retiarii.strategy.GridSearch`) samples new models from Model Space using a Grid Search algorithm.

Regularized Evolution

Regularized Evolution Strategy (`nni.retiarii.strategy.RegularizedEvolution`) implements Genetic Algorithm Search with mutation operator using Tournament Selection method. Regularized Evolution Strategy is close to Evolution Tuner we studied in Chapter 3. Pseudo-code that describes Regularized Evolution Algorithm is provided in the following.

Regularized Evolution Strategy has three global hyperparameters:

- `POPULATION_SIZE`: The size of a population that will try to participate in evolution search
- `CANDIDATES_N`: Number of candidates Tournament Selection method will pick from the population
- `GENERATION_N`: Total number of cycles

```
# HYPERPARAMETERS
POPULATION_SIZE
CANDIDATES_N
GENERATIONS_N
```

The population is initialized with random individuals (i.e., random neural architecture):

```
population = []
for _ in range(POPULATION_SIZE):
    individual = generate_random_architecture()
    evaluate(individual)
    population.append(individual)
```

```
for _ in range(GENERATIONS):
```

Algorithm picks CANDIDATES_N random individuals from population:

```
candidates = random_choice(population, CANDIDATES_N)
```

From these candidates, the best one is selected (i.e., the individual that has the best metric):

```
best_candidate = get_best_from(candidates)
```

Random mutation is performed on the best candidate (i.e., algorithm runs several Mutators in the original model):

```
mutant = mutate(best_candidate)
```

Mutant individual is being evaluated:

```
evaluate(mutant)
```

Mutant replaces the worst individual in the population:

```
replace_worst(population, mutant)
```

Figure 4-14 demonstrates the algorithm of Regularized Evolution Strategy.

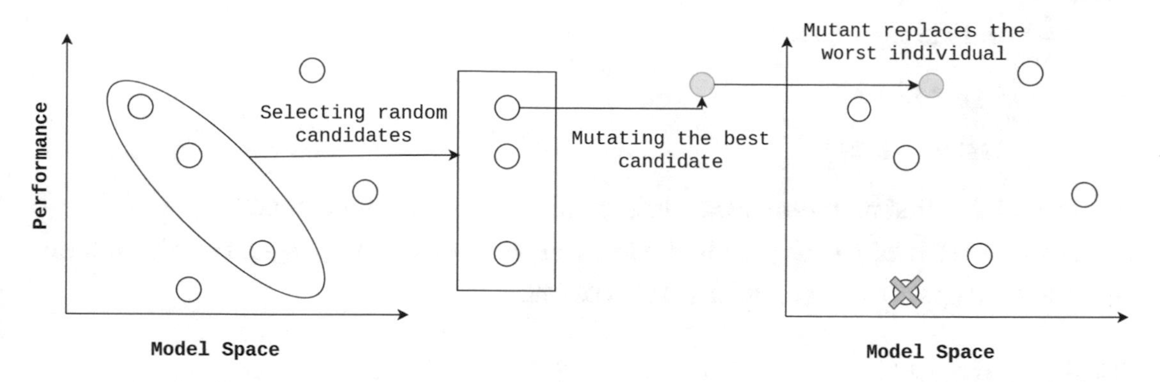

Figure 4-14. *Regularized Evolution Strategy*

The implementation of Regularized Evolution Strategy has the following parameters:

- `optimize_mode:`

  Type: `string`

  Default: `maximize`

  Values: '`maximize`' | '`minimize`'

  Sets the evolution direction.

- `population_size:`

  Type: `int`

  Default: `100`

  The number of individuals in the population.

- `cycles:`

  Type: `int`

  Default: `20000`

  The number of generations of the algorithm.

- `sample_size:`

  Type: `int`

  Default: `25`

  The number of individuals that should participate in each Tournament Selection.

- `mutation_prob`:

  Type: `float`

  Default: `0.05`

Probability that mutation occurs in each mutator in the Model Space.

For more details, please refer to the original paper describing Regularized Evolution approach: `https://arxiv.org/abs/1802.01548`.

TPE Strategy

TPE Strategy (`nni.retiarii.strategy.TPEStrategy`) is a Sequential Model-Based Optimization approach based on Tree-structured Parzen Estimator. It acts the same way as TPE Tuner we studied in Chapter 3.

For more details, please refer to the original paper describing the TPE approach: `https://papers.nips.cc/paper/2011/file/86e8f7ab32cfd12577bc2619bc635690-Paper.pdf`.

RL Strategy

RL Strategy (`nni.retiarii.strategy.PolicyBasedRL`) implements the Reinforcement Learning approach based on policy-gradient method (Proximal Policy Optimization or PPO). RL Strategy implements a special Recurrent Neural Network called Controller. Controller generates various model architectures from Model Space. The Controller acts as a stochastic policy; hence, it returns the mutation probability for each of the Mutators in the Model Space. After each trial, the Controller updates the weights of its RNN according to the Proximal Optimization Policy method. This approach allows exploring the Model Space by constructing a probability distribution for each of the mutators.

Figure 4-15 demonstrates RL Strategy in action.

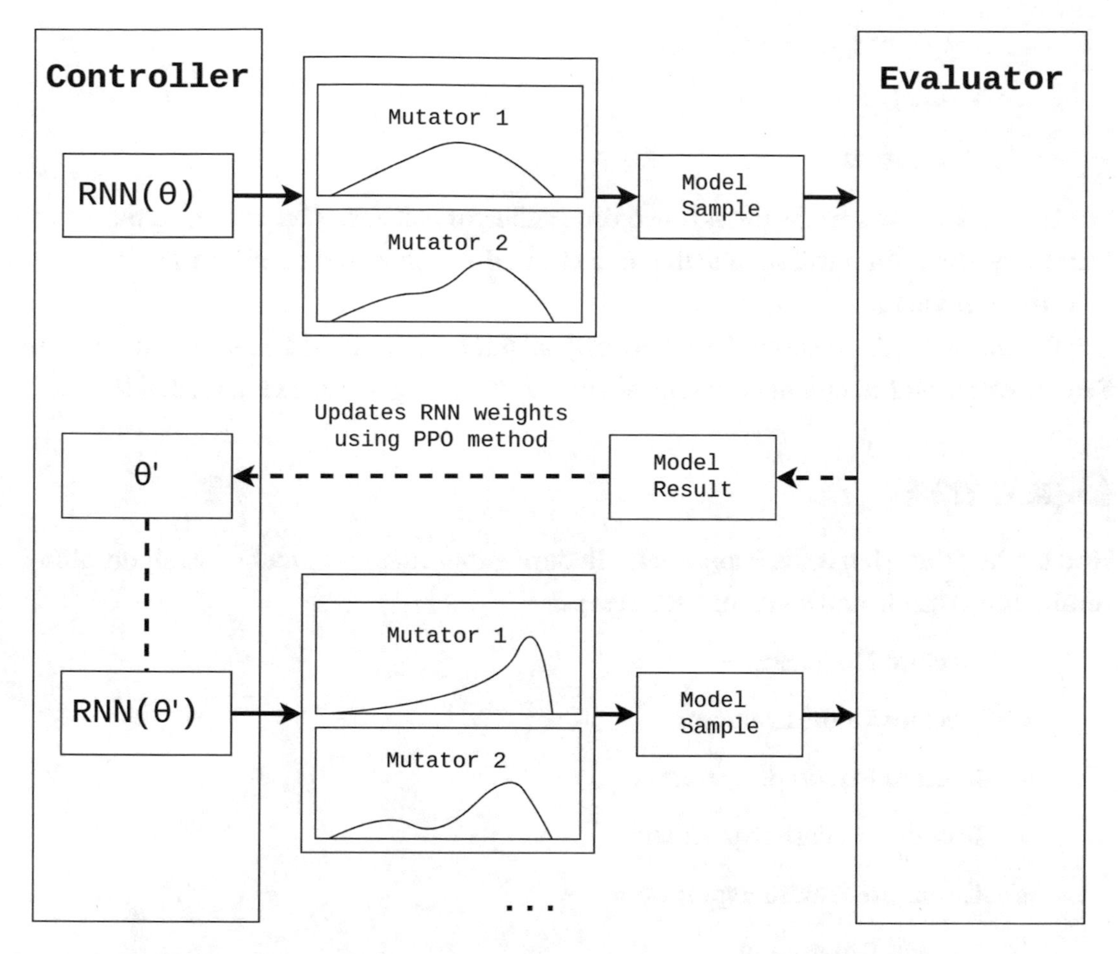

***Figure 4-15.** Reinforcement Learning (Proximal Policy Optimization) Strategy*

RL Strategy requires `tianshou` package to be installed. Tianshou is a Reinforcement Learning platform based on pure PyTorch.

```
pip install tianshou
```

The implementation of RL Strategy has the following parameters:

- `max_collect`:

  Type: `int`

  Default: `100`

  How many epochs Exploration Strategy performs.

- `Trial_per_collect`:

  Type: `int`

  Default: `20`

How many trials (trajectories) each time collector collects. After each completed trajectory, the trainer will sample the batch from the replay buffer and do the Controller update.

For more details, please refer to the original paper describing the Neural Architecture Search with Reinforcement Learning: `https://arxiv.org/pdf/1611.01578.pdf`.

Experiment

And the last thing left is the Experiment. Retiarii Experiment is launched in stand-alone (embedded) mode and contains seven steps:

- Declare Model Space
- Declare Model Evaluator
- Declare Exploration Strategy
- Initialize Retiarii Experiment
- Configure Retiarii Experiment
- Launch Experiment
- Returning results

The following pattern can be used to create Retiarii Experiment:

```
import nni.retiarii.strategy as strategy
from nni.retiarii import model_wrapper
from nni.retiarii.evaluator import FunctionalEvaluator
from nni.retiarii.experiment.pytorch import RetiariiExeConfig,
RetiariiExperiment,

# Declare Model Space
base_model = Net()

# Declare Model Evaluator
search_strategy = strategy.Random()
```

```
# Declare Exploration Strategy
model_evaluator = FunctionalEvaluator(evaluate_model)

# Initialize Retiarii Experiment
exp = RetiariiExperiment(base_model, model_evaluator, [], search_strategy)

# Configure Retiarii Experiment
exp_config = RetiariiExeConfig('local')
exp_config.experiment_name = 'mnist_search'
exp_config.trial_concurrency = 2
exp_config.max_trial_number = 20
exp_config.training_service.use_active_gpu = False
export_formatter = 'dict'

# uncomment this for graph-based execution engine
# exp_config.execution_engine = 'base'
# export_formatter = 'code'

# Launch Experiment
exp.run(exp_config, 8081 + random.randint(0, 100))

# Returning results
print('Final model:')
for model_code in exp.export_top_models(formatter=export_formatter):
    print(model_code)
```

You can use the WebUI after running the experiment the same way we did earlier launching HPO experiments.

CIFAR-10 LeNet NAS

Let's study the application of Multi-trial NAS to the CIFAR-10 problem. CIFAR-10 is a common dataset for image classification problem. It contains 60,000 color images (32×32 pixels) from 10 different classes: *airplane, automobile, bird, cat, deer, dog, frog, horse, ship,* and *truck.*

Please run the following command to download the CIFAR-10 dataset to your machine:

```
$ python3 ch4/utils/datasets.py
```

Figure 4-16 demonstrates several samples from CIFAR10 dataset.

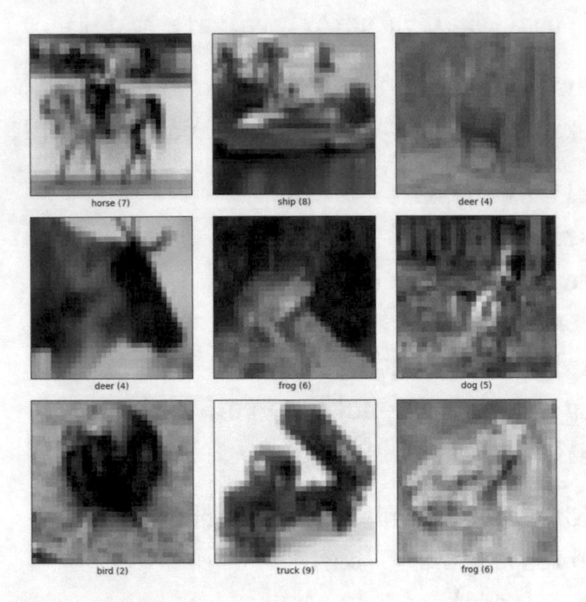

Figure 4-16. *CIFAR-10 samples*

Let's try to find an appropriate deep learning model based on the LeNet approach solving the CIFAR-10 classification problem. As we already know, the LeNet image recognition architecture can be divided into two components: *Feature Extraction Component* and *Decision Maker Component*. The *Feature Extraction Component* consists of a sequence of Feature Extraction blocks with convolution layer. *Decision Maker Component* consists of Fully Connected Components with linear layer.

The design of the Feature Extraction block can be as follows:

- `Conv` → `Activation`
- `Conv` → `Pool` → `Activation`

Figure 4-17 demonstrates possible architecture options for the Feature Extraction block or Feature Extraction block space.

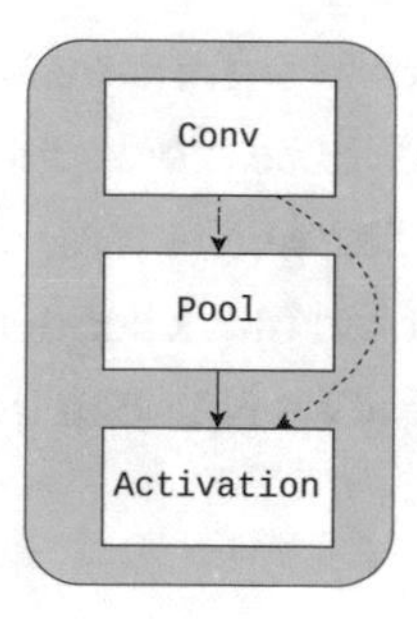

Figure 4-17. *Feature Extraction block space*

In the same way, we can determine the possible designs for the Fully Connected block:

- `Linear` → `Activation`
- `Linear` → `Dropout` → `Activation`

Figure 4-18 illustrates Fully Connected block space.

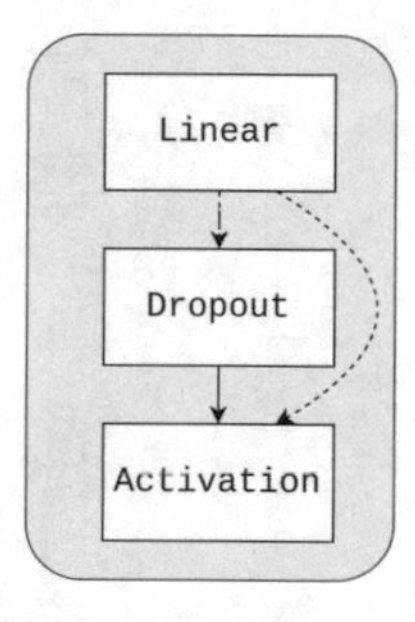

Figure 4-18. *Fully Connected block space*

Note Actually, `Linear` → `Dropout(p=1)` → `Activation` equals `Linear` → `Activation`, and it is possible to design the same block space from Figure 4-18 without using two connections. We could achieve the same block space using simple Layer Hyperparameter Optimization: `Linear` → `Dropout(p=[.3, .5, .8, 1])` → `Activation`. But we use two connections on purpose here because we want to demonstrate how the Multi-trial NAS chooses the best connection.

The neural architecture we are looking for consists of Feature Extraction and Fully Connected block sequences. Each of these blocks may have the architecture shown in Figures 4-17 and 4-18, respectively. The LeNet NAS algorithm must find the optimal Feature Extraction and Fully Connected block sequence lengths, as well as their architectures. The Model Space for LeNet NAS can be drawn as depicted in Figure 4-19.

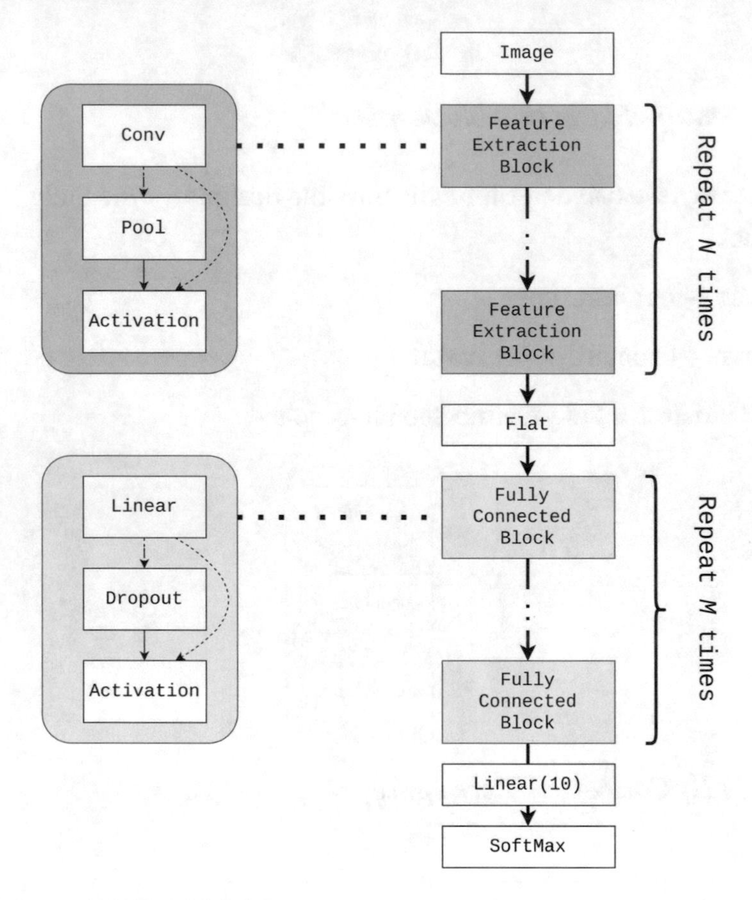

Figure 4-19. *LeNet NAS Model Space*

We will start implementing the CIFAR-10 LeNet NAS by defining a Feature Extraction block in Listing 4-7.

Listing 4-7. Feature Extraction block. ch4/retiarii/cifar_10_lenet/feature_extraction.py

```
from typing import Tuple
import nni.retiarii.nn.pytorch as nn
```

`FeatureExtractionBlock` takes the following parameters:

- `dim`: Input-output channels
- `kernel_size`: Kernel of convolution layer
- `activation`: Activation function
- `block_num`: Ordinal number of the block in Feature Extraction sequence

```
class FeatureExtractionBlock(nn.Module):

    def __init__(
            self,
            dim: Tuple[int, int],
            kernel_size,
            activation,
            block_num = 0
    ) -> None:

        super().__init__()
```

Initializing core convolution layer:

```
        self.input_dim = dim[0]
        self.output_dim = dim[1]

        self.conv = nn.Conv2d(
            in_channels = self.input_dim,
            out_channels = self.output_dim,
            kernel_size = kernel_size
        )
```

Declaring pool and activation layers:

```
        self.max_pool = nn.MaxPool2d(2, 2)
        self.activation = activation
```

Switch choice for two different branches:

- `Conv` → `Pool` → `Activation`
- `Conv` → `Activation`

```
self.switch = nn.InputChoice(
    n_candidates = 2,
    n_chosen = 1,
    label = f'fe_switch_{block_num}'
)
```

Defining forward method:

```
def forward(self, x):

    x = self.conv(x)

    # Branch A
    a = self.max_pool(x)
    a = self.activation(a)

    # Branch B
    b = self.activation(x)

    return self.switch([a, b])
```

The following method is used in `Repeat` Mutator. It returns an appropriate `FeatureExtractionBlock` by its ordinal number in the Feature Extraction sequence.

```
@classmethod
def create(cls, activation, in_dimension):

    def create_block(i):
        params = {
            'kernel_size': nn.ValueChoice(
                [3, 5],
                label = f'fe_kernel_size_{i}'
            )
        }
```

```
            # Feature Block Dimensions
            if i == 0:
                dim = (in_dimension, 8)
            elif i == 1:
                dim = (8, 16)
            else:
                dim = (16, 16)

            params['dim'] = dim
            params['activation'] = activation

            return FeatureExtractionBlock(**params)

        return create_block
```

Fully Connected block implementation is similar to Feature Extraction Block and is provided in Listing 4-8.

Listing 4-8. Fully Connected block. ch4/retiarii/cifar_10_lenet/fully_connected.py

```
from typing import Tuple, Iterator
from torch.nn import Parameter
import nni.retiarii.nn.pytorch as nn
```

`FullyConnectedBlock` takes the following parameters:

- `dim`: Input-output features for linear layer
- `dropout_rate`: Layer hyperparameter for dropout layer
- `activation`: Activation function
- `block_num`: Ordinal number of the block in Fully Connected sequence

```
class FullyConnectedBlock(nn.Module):

    def __init__(
            self,
            dim: Tuple[int, int],
```

```
        dropout_rate,
        activation,
        block_num
) -> None:

    super().__init__()
```

Initializing linear layer:

```
    self.input_dim = dim[0]
    self.output_dim = dim[1]
    self._linear = None
```

Declaring dropout and activation layers:

```
    self.dropout = nn.Dropout(p = dropout_rate)
    self.activation = activation
```

Switch choice for two different branches:

- `Linear → Dropout → Activation`
- `Linear → Activation`

```
    self.switch = nn.InputChoice(
        n_candidates = 2,
        n_chosen = 1,
        label = f'fc_switch_{block_num}'
    )
```

Defining forward method:

```
def forward(self, x):

    if not self.input_dim:
        self.input_dim = x.shape[1]

    # Branch A
    a = self.linear(x)
    a = self.dropout(a)
    a = self.activation(a)
```

```
        # Branch B
        b = self.linear(x)
        b = self.activation(b)

        return self.switch([a, b])
```

The following method is used in Repeat Mutator. It returns an appropriate FullyConnectedBlock by its ordinal number in the Fully Connected sequence.

```
    @classmethod
    def create(cls, activation, units, dropout_rate):

        def create_block(i):
            return FullyConnectedBlock(
                dim = (units[i], units[i + 1]),
                dropout_rate = dropout_rate,
                activation = activation,
                block_num = i
            )

        return create_block
```

And now we are ready to build LeNet Model Space.

Importing modules:

Listing 4-9. LeNet Model Space. ch4/retiarii/cifar_10_lenet/lenet_model_space.py

```
from typing import Iterator
from torch.nn import Parameter
import nni.retiarii.nn.pytorch as nn
from nni.retiarii import model_wrapper
from ch4.retiarii.cifar_10_lenet.feature_extraction import
FeatureExtractionBlock
from ch4.retiarii.cifar_10_lenet.fully_connected import FullyConnectedBlock
```

Do not forget to add @model_wrapper to the Model Space class:

```
@model_wrapper
class Cifar10LeNetModelSpace(nn.Module):

    def __init__(self):

        super().__init__()
        # number of classes for CIFAR-10 dataset
        self.class_num = 10

        # RGB input channels
        self.input_channels = 3
```

First, we define the space for the Feature Extraction sequence. All Feature Extraction blocks will share the same activation function:

```
        fe_activation = nn.LayerChoice(
            [nn.Sigmoid(), nn.ReLU()],
            label = f'fe_activation'
        )
```

Repeat mutator will create two or three Feature Extraction blocks in a row:

```
        self.fe = nn.Repeat(
            FeatureExtractionBlock.create(fe_activation, self.input_channels),
            depth = (2, 3), label = 'fe_repeat'
        )
```

Second, we define a Fully Connected sequence:

```
        self.flat = nn.Flatten()
```

All Fully Connected blocks will share the same activation function:

```
        dm_activation = nn.LayerChoice(
            [nn.Sigmoid(), nn.ReLU()],
            label = f'fc_activation'
        )
```

Layer hyperparameters for Fully Connected blocks:

```
l1_size = nn.ValueChoice([256, 128], label = 'l1_size')
l2_size = nn.ValueChoice([128, 64], label = 'l2_size')
l3_size = nn.ValueChoice([64, 32], label = 'l3_size')
dropout_rate = nn.ValueChoice([.3, .5], label = 'fc_dropout_rate')
```

Repeat mutator will create from one to three Fully Connected blocks in a row:

```
self.dm = nn.Repeat(
    FullyConnectedBlock.create(
        dm_activation,
        [None, l1_size, l2_size, l3_size],
        dropout_rate
    ),
    depth = (1, 3), label = 'fc_repeat'
)
```

Final pair of classification layers (`linear_final` layer initialized lazily):

```
self.linear_final_input_dim = None
self._linear_final = None
self.log_max = nn.LogSoftmax(dim = 1)
```

Executing `forward` method:

```
def forward(self, x):
    x = self.fe(x)
    x = self.flat(x)
    x = self.dm(x)

    if not self.linear_final_input_dim:
        self.linear_final_input_dim = x.shape[1]

    x = self.linear_final(x)
    return self.log_max(x)
```

Model evaluator is a classical neural network train–test algorithm. You can examine its code here: ch4/retiarii/cifar_10_lenet/eval.py.

Fine! Since LeNet Model Space is ready, we can start the research with the code in Listing 4-10.

Importing modules:

Listing 4-10. LeNet NAS Experiment. ch4/retiarii/cifar_10_lenet/run_cifar10_lenet_experiment.py

```
from nni.retiarii.evaluator import FunctionalEvaluator
from nni.retiarii.experiment.pytorch import RetiariiExperiment,
RetiariiExeConfig
import nni.retiarii.strategy as strategy
from ch4.retiarii.cifar_10_lenet.eval import evaluate
from ch4.retiarii.cifar_10_lenet.lenet_model_space import
Cifar10LeNetModelSpace
```

Declaring Model Space:

```
model_space = Cifar10LeNetModelSpace()
```

Defining Model Evaluator:

```
evaluator = FunctionalEvaluator(evaluate)
```

We will use RL Search Strategy for this experiment:

```
search_strategy = strategy.PolicyBasedRL(
    trial_per_collect = 10,
    max_collect = 200
)
```

Initializing Retiarii Experiment:

```
exp = RetiariiExperiment(model_space, evaluator, [], search_strategy)
```

Experiment configuration:

```
exp_config = RetiariiExeConfig('local')
exp_config.experiment_name = 'CIFAR10_LeNet_NAS'
exp_config.trial_concurrency = 1
exp_config.max_trial_number = 500
exp_config.training_service.use_active_gpu = False
export_formatter = 'dict'
```

Launching Experiment:

```
exp.run(exp_config, 8080)
```

Returning results:

```
print('Final model:')
for model_code in exp.export_top_models(formatter = export_formatter):
    print(model_code)
```

The experiment can be run as follows:

```
$ python3 ch4/retiarii/cifar_10_lenet/run_cifar10_lenet_experiment.py
```

Note Duration ~ 20 hours on Intel Core i7 with CUDA (GeForce GTX 1050)

The best model shows **0.84** accuracy on test dataset. It is not a bad result, but it seems there is still room for improvement. The best model has the following parameters:

```
{
  "fe_repeat": 2,
  "fe_kernel_size_0": 3,
  "fe_activation": "1",
  "fe_switch_0": 0,
  "fe_kernel_size_1": 5,
  "fe_kernel_size_2": 3,
  "fc_repeat": 2,
  "l1_size": 128,
  "fc_dropout_rate": 0.3,
  "fc_switch_0": 0,
  "l2_size": 64,
  "fc_switch_1": 0,
  "l3_size": 64,
  "fc_switch_2": 0
}
```

The preceding parameters can be interpreted the following way:

`"fe_repeat": 2` – means that `Repeat` mutator generates the Feature Extraction sequence of two blocks:

```
self.fe = nn.Sequential(
    [
        FeatureExtractionBlock.create(fe_activation, self.input_
        channels)(0),
        FeatureExtractionBlock.create(fe_activation, self.input_
        channels)(1),
    ]
)
```

`"fe_kernel_size_0": 3` – means that `3` is chosen in `ValueChoice` mutator:

```
'kernel_size': nn.ValueChoice(
    [3, # <- this value
     5],
    label = f'fe_kernel_size_{i}'
)
```

`"fe_switch_0": 0` – means that the first connection is used in `InputChoice` mutator:

```
self.switch([a, b]) <- returns a
```

Also, it is convenient to visualize the architecture using Netron in Trial details panel. Figure 4-20 demonstrates the architecture of the best model.

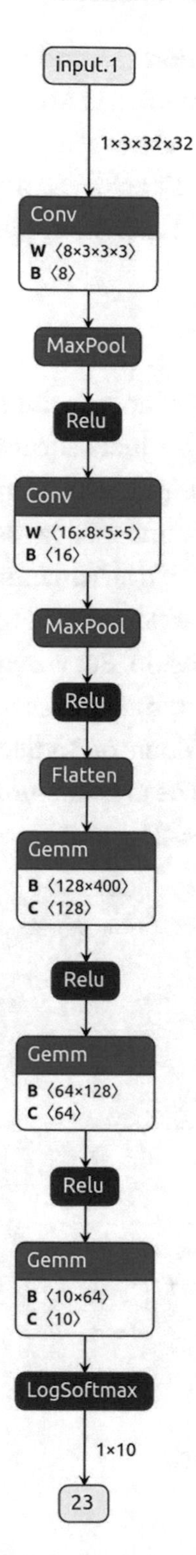

Figure 4-20. *LeNet NAS Best Model Architecture*

The research we made in this section can be a good starting point for your own NAS solutions. It contains basic techniques used in Multi-trial NAS. Here, we used the LeNet model as the Base Model, but you can choose any model and any mutators that fit the concrete problem better. But we haven't achieved a great result in the LeNet NAS. In the next section, we'll try another NAS with a more sophisticated approach.

CIFAR-10 ResNet NAS

Let's try to find another approach to the architecture search that solves the CIFAR-10 problem. In 2015, the article "Deep Residual Learning for Image Recognition" (`https://arxiv.org/pdf/1512.03385.pdf`) was published. This article had a huge impact on deep learning in general. It introduced the concept of a residual term, which could degrade the blocks of the neural network, which overwhelmed its performance.

The basic building block of the ResNet model is the Bottleneck block. Original Bottleneck block skips connections inside. But we will add this as an optional feature. NAS algorithm will define if it is appropriate to use the skip connection technique for the Bottleneck block. So there are two variants of Bottleneck block in ResNet NAS *with* and *without* skip connection technique. The architecture of the Bottleneck block space can be demonstrated as shown in Figure 4-21.

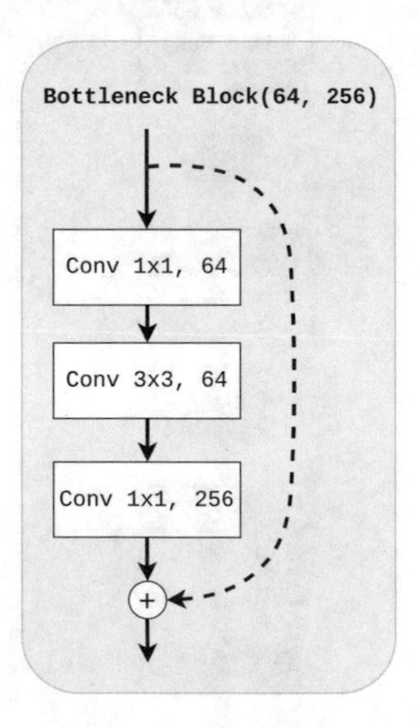

Figure 4-21. *Bottleneck block space*

Another component of ResNet is the Residual Cell. Residual Cell consists of a sequence of Bottleneck blocks. The optimal length of the Bottleneck sequence in Residual Cell depends on the concrete dataset. Figure 4-22 shows the Residual Cell space.

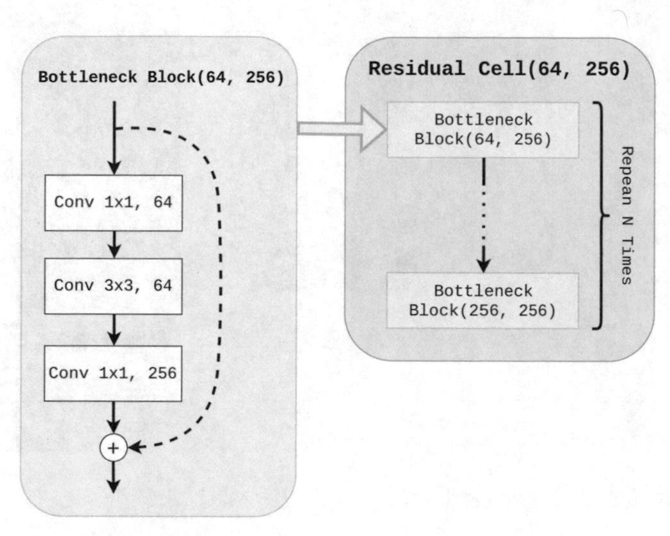

Figure 4-22. *Residual Cell space*

And finally, we can construct the ResNet model. ResNet has the following architecture:

- Initial convolution layer
- Residual Cell sequence
- Several Fully Connected layers

The optimal length of a Residual Cell sequence depends on the dataset. And we will try to find it in ResNet NAS. Figure 4-23 presents the complete Model Space for ResNet NAS.

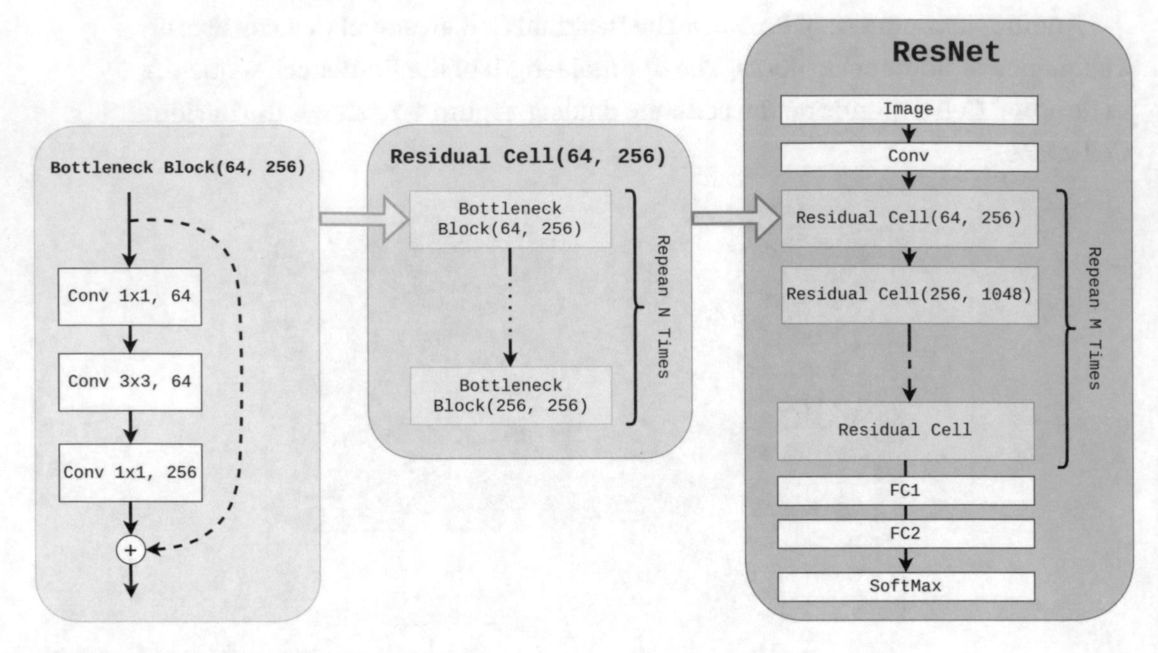

Figure 4-23. *ResNet NAS space*

Listing 4-11 defines the Bottleneck block for ResNet Model Space.

Listing 4-11. ResNet NAS Bottleneck block. ch4/retiarii/cifar_10_resnet/bottle_neck.py

```
import nni.retiarii.nn.pytorch as nn

class Bottleneck(nn.Module):
```

Bottleneck expansion ratio:

```
    expansion = 4
```

The number of output channels that are produced by Bottleneck block:

```
    @classmethod
    def result_channels_num(cls, channels):
        return channels * cls.expansion
```

Bottleneck takes the following parameters:

- `cell_num`: The ordinal number of Residual Cell this block belongs to
- `in_channels`: Input channels to the first convolution layer

- `out_channels`: Output channels to all convolutional layers in the block
- `i_downsample`: Identity downsample block

```
def __init__(
        self,
        cell_num,
        in_channels,
        out_channels,
        i_downsample = None,
        stride = 1
):
    super(Bottleneck, self).__init__()
```

Defining three convolution layers with batch normalization:

```
self.conv1 = nn.Conv2d(
    in_channels, out_channels,
    kernel_size = 1, stride = 1, padding = 0
)
self.batch_norm1 = nn.BatchNorm2d(out_channels)

self.conv2 = nn.Conv2d(
    out_channels, out_channels,
    kernel_size = 3, stride = stride, padding = 1
)
self.batch_norm2 = nn.BatchNorm2d(out_channels)

self.conv3 = nn.Conv2d(
    out_channels, self.result_channels_num(out_channels),
    kernel_size = 1, stride = 1, padding = 0
)
self.batch_norm3 = nn.BatchNorm2d(self.result_channels_num(out_
channels))
```

Skip connection acts the same for all blocks in Residual Cell because all `InputChoice` mutators share the same label in the Residual Cell:

```
self.skip_connection = nn.InputChoice(
    n_candidates = 2,
    n_chosen = 1,
    label = f'bottle_neck_{cell_num}_skip_connection'
)
```

Identity downsampling block:

```
self.i_downsample = i_downsample

self.stride = stride
self.relu = nn.ReLU()
```

Defining forward method:

```
def forward(self, x):

    # x0
    identity = x.clone()

    x = self.relu(self.batch_norm1(self.conv1(x)))

    x = self.relu(self.batch_norm2(self.conv2(x)))

    x = self.conv3(x)
    x = self.batch_norm3(x)

    identity = self.skip_connection([identity, None])
```

Skipping connection if `self.skip_connection` returns not `None`:

```
if identity is not None:
    #downsample if needed
    if self.i_downsample is not None:
        identity = self.i_downsample(identity)
    # adding identity
    x += identity
```

```
        x = self.relu(x)

        return x
```

Since we have defined the Bottleneck block, we can move to the ResNet Model Space definition in Listing 4-12.

(Some unimportant code segments are omitted. Complete code is provided in the corresponding file: *ch4/retiarii/cifar_10_resnet/res_net_model_space.py*.)

Importing modules:

Listing 4-12. ResNet NAS Model Space

```
from typing import Iterator
import nni.retiarii.nn.pytorch as nn
from nni.retiarii import model_wrapper
from torch.nn import Parameter
from ch4.retiarii.cifar_10_resnet.bottle_neck import Bottleneck
```

Don't forget to annotate Model Space with `@model_wrapper`:

```
@model_wrapper
class ResNetModelSpace(nn.Module):
```

Global model constants:

```
    # classification classes
    num_classes = 10
    # input channels for RGB image
    in_channels = 3
    # ResNet Channel constant
    channels = 64

    def __init__(self):
        super().__init__()
```

Choosing `ReLU` as a global activation function:

```
        self.relu = nn.ReLU()
```

Entry point convolution layer with batch normalization:

```
self.conv1 = nn.Conv2d(
    in_channels = self.in_channels,
    out_channels = self.channels,
    kernel_size = 7,
    stride = 2,
    padding = 3,
    bias = False
)

self.batch_norm1 = nn.BatchNorm2d(64)
```

MaxPool layer with the following hyperparameter list `[2, 3]`:

```
pool_size = nn.ValueChoice([2, 3], label = 'pool_size')
self.max_pool = nn.MaxPool2d(kernel_size = pool_size, stride = 2,
padding = 1)
```

Constructing Residual Cell sequence with `Repeat` mutator (from two to five cells):

```
self.res_cells = nn.Repeat(
    ResNetModelSpace.residual_cell_builder(),
    depth = (2, 5), label = 'res_cells_repeat'
)
```

Constructing Fully Connected sequence with two linear layers:

```
self.avg_pool = nn.AdaptiveAvgPool2d((1, 1))
self.fc1_input_dim = None
self.fc1_output_dim = nn.ValueChoice(
    [256, 512],
    label = 'fc1_output_dim'
)
self._fc1 = None
self.fc2 = nn.Linear(
    in_features = self.fc1_output_dim,
    out_features = self.num_classes
)
```

Defining `forward` method:

```
def forward(self, x):
    x = self.relu(self.batch_norm1(self.conv1(x)))
    x = self.max_pool(x)

    x = self.res_cells(x)

    x = self.avg_pool(x)
    x = x.reshape(x.shape[0], -1)

    if not self.fc1_input_dim:
        self.fc1_input_dim = x.shape[1]

    x = self.relu(self.fc1(x))
    x = self.fc2(x)
    return x
```

The following method is used to construct Residual Cells for the `Repeat` mutator:

```
@classmethod
def residual_cell_builder(cls):

    def create_cell(cell_num):
```

Defining Residual Cell parameters:

```
        # planes sequence: 64, 128, 256, 512,...
        planes = 64 * pow(2, cell_num)
        # stride sequence: 1, 2, 2, 2,...
        stride = max(1 + cell_num, 2)
```

Number of Bottleneck blocks in the Residual Cell:

```
        # block sequence: 3, 4, 5, 5,...
        blocks = max(3 + cell_num, 5)

        downsample = None
```

Placeholder for Bottleneck blocks:

```
        layers = []
```

Constructing downsample identity block if needed:

```
if stride != 1 or cls.channels != Bottleneck.result_channels_
num(planes):
    downsample = nn.Sequential(
        nn.Conv2d(
            in_channels = cls.channels,
            out_channels = Bottleneck.result_channels_
            num(planes),
            kernel_size = 1,
            stride = stride
        ),
        nn.BatchNorm2d(
            num_features = Bottleneck.result_channels_num(planes)
        )
    )

layers.append(
    Bottleneck(
        cell_num = cell_num,
        in_channels = cls.channels,
        out_channels = planes,
        i_downsample = downsample,
        stride = stride
    )
)

cls.channels = Bottleneck.result_channels_num(planes)
```

Generating sequence of Bottleneck blocks:

```
for i in range(blocks - 1):
    layers.append(
        Bottleneck(
            cell_num = cell_num,
            in_channels = cls.channels,
            out_channels = planes
        )
```

```
            )

        return nn.Sequential(*layers)

    return create_cell
```

Phew. It was not easy to follow the definition of ResNet Model Space if you are not familiar with ResNet yet. Anyway, don't forget that NAS treats the neural network as Data Flow Graph, and it tries to find the optimal combination of nodes and connections in the Model Space we constructed in Figure 4-23. Even if some deep learning concepts are not familiar to you yet, try to treat the NAS as the search for the optimal subgraph in supergraph space.

ResNet model evaluator is a classical neural network train–test algorithm. You can examine its code here: ch4/retiarii/cifar_10_resnet/eval.py. ResNet NAS Experiment script does not differ too much from LeNet NAS, and I don't provide its code in the book. Please refer to the script file: ch4/retiarii/cifar_10_resnet/run_cifar10_resnet_experiment.py.

The experiment can be run as follows:

```
$ python3 ch4/retiarii/cifar_10_resnet/run_cifar10_resnet_experiment.py
```

Note Duration ~ 60 hours on Intel Core i7 with CUDA (GeForce GTX 1050)

The best model shows **0.957** accuracy on test dataset. That is not perfect, but a much better result than LeNet NAS produced (**0.84**). The best neural network architecture has the following parameters:

```
{
  "pool_size": 2,
  "res_cells_repeat": 5,
  "bottle_neck_0_skip_connection": 1,
  "bottle_neck_1_skip_connection": 0,
  "bottle_neck_2_skip_connection": 0,
  "bottle_neck_3_skip_connection": 0,
  "bottle_neck_4_skip_connection": 0,
  "fc1_output_dim": 512
}
```

These parameters mean that the best model has the sequence of five Residual Cells: the first cell is not using the skipping connection technique, and the others use it. This is a predictable result because usually skipping connection technique raises neural network performance.

In this section, we have achieved a great result! CIFAR-10 is a highly complex computer vision problem, and even large, sophisticated neural networks cannot reach high accuracy with this dataset. Table 4-2 compares the architecture we have constructed in this section with other common architectures.

Table 4-2. *Rating of the best architectures for CIFAR-10*

Rank	Architecture	Accuracy
79	AutoDropout	96.8
96	Wide ResNet	96.11
-	**Multi-trial NAS ResNet Result**	**95.7**
104	SimpleNetv1	95.51
108	MomentumNet	95.18
116	VGG-19 with GradInit	94.71
128	Tree+Max-Avg pooling	94

For more detailed information concerning architecture performance on the CIFAR-10 problem, please refer to `https://paperswithcode.com/sota/image-classification-on-cifar-10`.

Classic Neural Architecture Search (TensorFlow)

Classic NAS implements the Multi-trial NAS approach. Evaluator takes a model from Model Space and evaluates it separately. Last valid documentation regarding Classic NAS with NNI can be found here: `https://nni.readthedocs.io/en/v2.2/nas.html`. NNI currently supports Classic NAS but is deprecating it in favor of the Retiarii framework. The procedure of Classic NAS algorithms is similar to hyperparameter tuning. Users use `nnictl` to start experiments, and each model runs as a trial. The difference is that the search space file is automatically generated from the Model Space by running `nnictl ss_gen`.

The main logical entities of Classic NAS are *Base Model, Mutator, Search Space, Trial,* and *Search Strategy*. Let's study each of them by applying NAS algorithm to the classical MNIST problem.

Base Model

Every Neural Architecture Search starts with defining a Base Model. The Base Model is a neural network that acts as a starting point for new architectures. The Base Model can be very simple or very complex. The researcher chooses the model that is most suitable as a baseline.

Let's examine the classic LeNet model for the MNIST problem in Listing 4-13.

Listing 4-13. Base Model. ch4/classic/base_model.py

```
class LeNetModel(Model):
```

Base Model layers:

```
    def __init__(self):
        super().__init__()
        self.conv1 = Conv2D(6, 3, padding = 'same', activation = 'relu')
        self.pool = MaxPool2D(2)
        self.conv2 = Conv2D(16, 3, padding = 'same', activation = 'relu')

        self.bn = BatchNormalization()

        self.gap = AveragePooling2D(2)
        self.fc1 = Dense(120, activation = 'relu')
        self.fc2 = Dense(84, activation = 'relu')
        self.fc3 = Dense(10)
```

Feed forward:

```
    def call(self, x):
        batch_size = x.shape[0]

        x = self.conv1(x)
        x = self.pool(x)
        x = self.conv2(x)
```

```
        x = self.pool(self.bn(x))
        x = self.gap(x)
        x = tf.reshape(x, [batch_size, -1])
        x = self.fc1(x)
        x = self.fc2(x)
        x = self.fc3(x)
        return x
```

The model described in Listing 4-13 will serve as a piece of clay for new architectures.

Mutators

Mutator transforms the Base Model into a new one. A set of mutators allow defining a search space for the NAS. Classic NNI NAS provides two mutators: `LayerChoice` and `InputChoice`. `LayerChoice` mutator forms the candidate layers for a layer placeholder. One of the candidates is tried in the exploration process. `InputChoice` tries different connections. It takes several tensors and chooses `n_chosen` tensors from them.

You can learn more about `LayerChoice` and `InputChoice` mutators in the subsection "Mutators" under the section "Neural Architecture Search Using Retiarii (PyTorch)." We can apply `LayerChoice` and `InputChoice` mutators to the Base Model in the following way.

Listing 4-14. LeNet Model Space. ch4/classic/model.py

```
class LeNetModelSpace(Model):

    def __init__(self):
        super().__init__()
```

We try three different convolution layers for the `conv1` placeholder:

```
        self.conv1 = LayerChoice([
            Conv2D(6, 3, padding = 'same', activation = 'relu'),
            Conv2D(6, 5, padding = 'same', activation = 'relu'),
            Conv2D(6, 7, padding = 'same', activation = 'relu'),
        ], key = 'conv1')
```

We try two different pooling layers for the pool placeholder:

```
self.pool = LayerChoice([
    MaxPool2D(2),
    MaxPool2D(3)],
    key = 'pool'
)
```

We try three different convolution layers for the conv2 placeholder:

```
self.conv2 = LayerChoice([
    Conv2D(16, 3, padding = 'same', activation = 'relu'),
    Conv2D(16, 5, padding = 'same', activation = 'relu'),
    Conv2D(16, 7, padding = 'same', activation = 'relu'),
], key = 'conv2')
self.conv3 = Conv2D(16, 1)
```

We add skip connection technique:

```
self.skip_connect = InputChoice(
    n_candidates = 2,
    n_chosen = 1,
    key = 'skip_connect'
)
self.bn = BatchNormalization()

self.gap = AveragePooling2D(2)
self.fc1 = Dense(120, activation = 'relu')
```

We add two candidates for fc1 placeholder:

```
self.fc2 = LayerChoice([
    Dense(84, activation = 'relu'),
    Layer()
], key = 'fc2')
self.fc3 = Dense(10)
```

Figure 4-24 demonstrates the set of all possible architectures described by Listing 4-14.

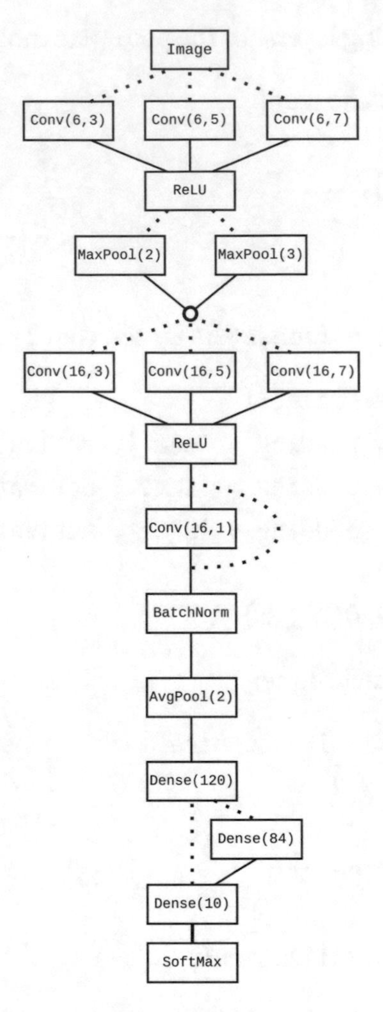

Figure 4-24. *Set of all possible architectures*

We see that each mutator adds variance to the Base Model.

Trial

NAS Trial means the same as the Trial in the HPO context. It initializes the model, trains it, tests it, and returns model accuracy. There is only one new feature: you must use the `get_and_apply_next_architecture` method from `nn.algorithms.nas.tensorflow.classic_nas` package to initialize the model. Listing 4-15 provides the NAS Trial.

(Full code is provided in the corresponding file: *ch4/classic/trial.py*.)

Listing 4-15. NAS Trial.

```
net = LeNetModelSpace()
get_and_apply_next_architecture(net)
train_model(net, dataset_train, optimizer, epochs)
acc = test_model(net, dataset_test)
nni.report_final_result(acc.numpy())
```

Search Space

After the Trial script is defined, you should generate the search space JSON file manually using the following command:

```
$ nnictl ss_gen --trial_command="python3 trial.py" --trial_dir=ch4/classic
--file=ch4/classic/search_space.json
```

The preceding command generates the file shown in Listing 4-16.

Listing 4-16. NAS search space

```
{
  "conv1": {
    "_type": "layer_choice",
    "_value": ["0", "1", "2"]
  },
  "conv2": {
    "_type": "layer_choice",
    "_value": ["0", "1", "2"]
  },
  "fc2": {
    "_type": "layer_choice",
    "_value": ["0", "1"]
  },
  "pool": {
    "_type": "layer_choice",
    "_value": ["0", "1"]
  },
```

```
  "skip_connect": {
    "_type": "input_choice",
    "_value": {
      "candidates": ["",""],
      "n_chosen": 1
    }
  }
}
```

As you can see, NNI Classic NAS implementation is pretty close to HPO implementation. The search space file is a list of all possible neural architecture choices.

Search Strategy

Classic NAS supports the following Search Strategies:

- `Random Search`
- `PPO Tuner`: Reinforcement Learning Tuner based on Proximal Policy Optimization algorithm

For more information about these tuners, you can refer to the subsection "Exploration Strategies" under the section "Neural Architecture Search Using Retiarii (PyTorch)."

Experiment

The last thing left to do is to define the experiment configuration. Experiment configuration is defined in Listing 4-17.

Listing 4-17. NAS configuration file

```
experimentName: example_mnist
trialConcurrency: 1
maxTrialNum: 100
trainingServicePlatform: local
searchSpacePath: search_space.json
```

```
tuner:
  builtinTunerName: PPOTuner
  classArgs:
    optimize_mode: maximize
trial:
  command: python3 trial.py
```

And now, we can start the experiment with the following command:

```
$ nnictl create --config=ch4/classic/config.yml
```

Note Duration ~ 2 hours on Intel Core i7 with CUDA (GeForce GTX 1050)

Experiment returns the best accuracy **0.9923** on test dataset, with the following parameter set:

```
conv1: 2
pool: 0
conv2: 0
fc2: 0
skip_connect: 0
```

The preceding parameters can be interpreted as the following neural architecture:

```
self.conv1 = LayerChoice([
    Conv2D(6, 3, ...),
    Conv2D(6, 5, ...),
    Conv2D(6, 7, ...), # <- 2
], key = 'conv1')

self.pool = LayerChoice([
    MaxPool2D(2), # <- 0
    MaxPool2D(3)],
    key = 'pool'
)
```

```
self.conv2 = LayerChoice([
    Conv2D(16, 3, ...), # <- 0
    Conv2D(16, 5, ...),
    Conv2D(16, 7, ...),
], key = 'conv2')

self.fc2 = LayerChoice([
    Dense(84, activation = 'relu'), # <- 0
    Layer()
], key = 'fc2')

x0 = self.skip_connect([
    x0, # <- 0
    None]
)
```

The best neural architecture is shown in Figure 4-25.

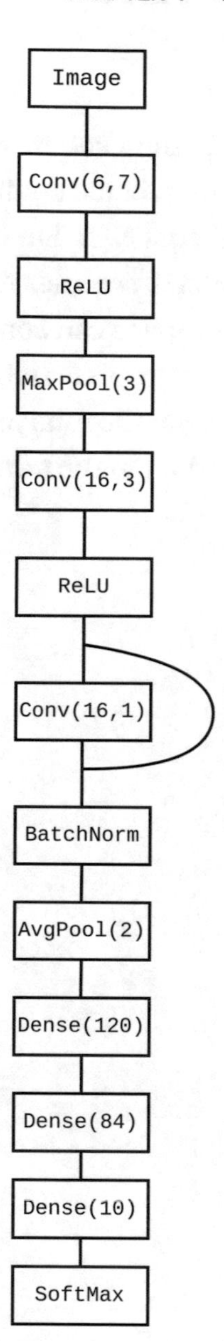

Figure 4-25. *Best neural architecture*

Classic NAS does not differ too much from the HPO approach. Indeed, we could build the same experiment with layer and design hyperparameter search. We made a close trick in Chapter 2, in the section "From LeNet to AlexNet." Neural Architecture Search has made great strides lately, and Classic NAS cannot support new research ideas. In any case, you can still use Classic NAS and get meaningful results by searching for new solutions.

Summary

Multi-trial Neural Architecture Search using Retiarii and classic approaches offers clear and elegant solutions for searching robust neural architectures. Many meaningful results could be achieved using Multi-trial NAS. But this approach has one very serious drawback. It takes too much time. Indeed, complex models and massive datasets need too much time to train, and the Model Space can contain millions of model samples. Even the most advanced Exploration Strategy can take too much time to converge to some suboptimal neural architecture. But the time problem has a solution called One-shot NAS, and we will explore this method in the next chapter.

CHAPTER 5

One-Shot Neural Architecture Search

In the previous chapter, we explored Multi-trial Neural Architecture Search, which is a very promising approach. And the reader might wonder why Multi-trial NAS is called that way. Are there any other non-Multi-trial NAS approaches, and is it really possible to search for the optimal neural network architecture in some other way without trying it? It looks pretty natural that the only way to find the optimal solution is to try different elements in the search space. In fact, it turns out that this is not entirely true. There is an approach that allows you to find the best architecture by training some Supernet. And this approach is called **One-shot Neural Architecture Search**. As the name "one-shot" implies, this approach involves only one try or shot. Of course, this "shot" is much longer than single neural network training, but nevertheless, it saves a lot of time.

In this chapter, we will study what One-shot NAS is and how to design architectures for this approach. We will examine two popular One-shot algorithms: **Efficient Neural Architecture Search via Parameter Sharing** (**ENAS**) and **Differentiable Architecture Search** (**DARTS**). Of course, we will apply these algorithms to solve practical problems.

`NNI 2.7` version (which is used in this book) has ENAS algorithm implementation for the TensorFlow framework, but it doesn't have one for the DARTS algorithm. Anyway, ENAS algorithm is one of the most popular and efficient One-shot NAS implementations, so TensorFlow users shouldn't get too frustrated.

One-Shot NAS in Action

Interest in automating neural network architecture design is growing, but the classical Multi-trial NAS approach is too computationally expensive, requiring thousands and millions of different architectures to be trained from scratch. This fact, unfortunately,

I. Gridin, *Automated Deep Learning Using Neural Network Intelligence*,
https://doi.org/10.1007/978-1-4842-8149-9_5

makes Multi-trial NAS not applicable in practice. Indeed, some Multi-trial experiments can take weeks or months on the most modern computing resources. A new One-shot NAS approach has been proposed to address this weakness of Multi-trial architecture search.

The best way to introduce One-shot NAS is to provide an example. Let's say we are looking for the optimal architecture for the MNIST problem. And we have the Model Space shown in Figure 5-1.

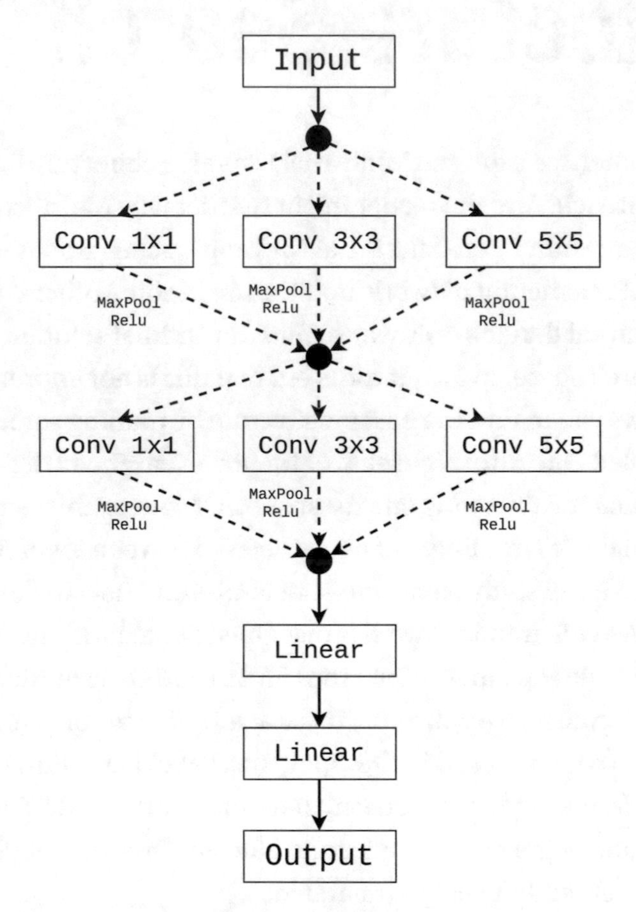

Figure 5-1. *Model Space for MNIST problem*

Figure 5-1 shows the Model Space with two mutable layers. Each mutable layer has the following choices: `Conv 1×1`, `Conv 3×3`, and `Conv 5×5`. In a classic Multi-shot scenario, we would perform 3×3=9 trials for each combination of parameters and pick the best one. While the One-shot NAS approach follows another technique, we create one *Supernet* that merges or reduces the output of each mutable layer and train the resulting neural network only once. Figure 5-2 demonstrates this Supernet.

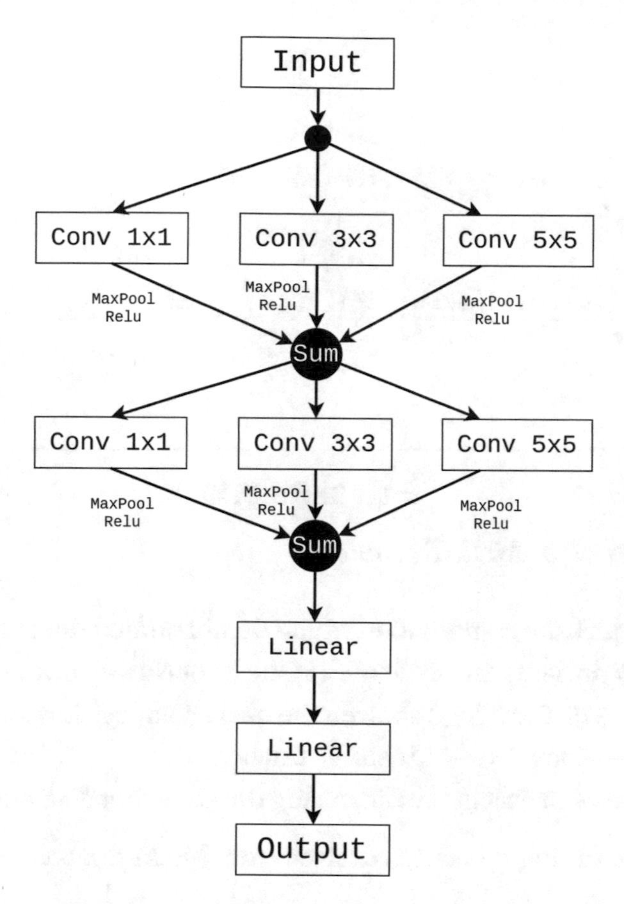

Figure 5-2. *Supernet for MNIST problem*

After training the Supernet, we evaluate it by activating each combination of layers and zeroing out other layers. Figure 5-3 illustrates this concept.

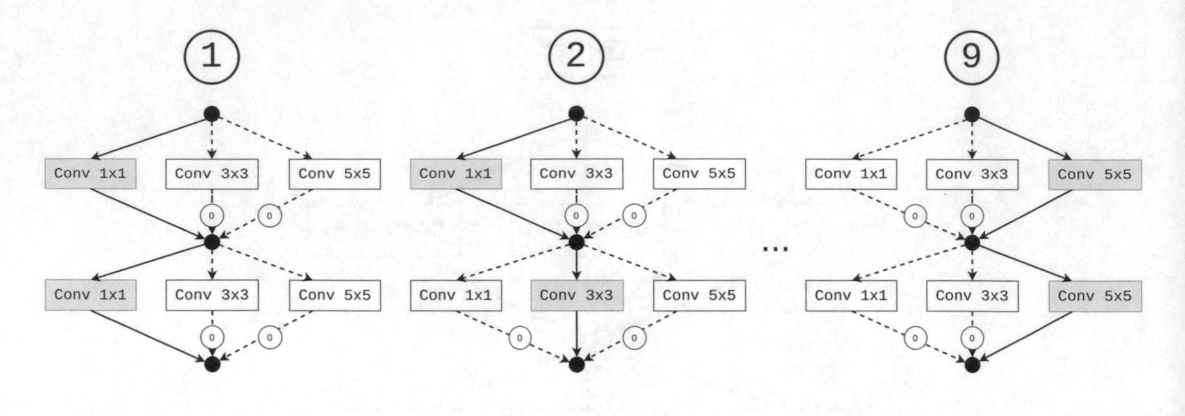

Figure 5-3. *Supernet for MNIST problem*

And finally, we pick the combination which demonstrated the best performance. This combination represents the result of the One-shot NAS algorithm. For example, if a combination (`Conv 5×5`, `Conv 5×5`) showed the best accuracy, then our target network design is `Conv 5×5` → `Conv 5×5` → `Linear` → `Linear`.

Let's summarize what exactly we did during the One-shot NAS algorithm:

- We created a single *Supernet* derived from Model Space.
- We *trained* it.
- We *evaluated* it nine times, activating different layers in turn.
- We picked the *best* neural design for the problem.

The main benefit we have here is that we trained Supernet only once instead of training each of nine candidate networks! It speeds up the whole neural architecture search dramatically, because the network training is the longest part of NAS process.

A reader may have a fair question: *"But wait! We trained a single Supernet network. All candidate layers learned to work together! But then we decided to break it into different parts, leaving the same weights. This is nonsense!"* I agree. This is a very counterintuitive concept. Indeed, all layers were trained together, and they learned to complement and help each other in solving the problem. Surely you can't just throw out some layers from a neural network searching for the best architecture. But the most fantastic thing about One-shot NAS is that you can! There is still no sufficient mathematical basis for this approach, but it works in practice. Let's implement this approach in practice using the example we considered earlier. In this section, we will not be using the NNI toolkit. Here, our goal is to get an intuition about the One-shot NAS approach.

To begin with, we will make a vanilla Multi-trial NAS. Listing 5-1 (TensorFlow implementation) and Listing 5-2 (PyTorch implementation) implement the model shown in Figure 5-1.

We import necessary modules:

Listing 5-1. **TensorFlow** implementation. Model Space. ch5/naive_one_shot_nas/tf/tf_lenet_multi_model.py

```
from tensorflow.keras import Model
from tensorflow.keras.layers import Dense, Flatten, MaxPool2D
from ch5.naive_one_shot_nas.tf.tf_ops import create_conv
```

The following model accepts two parameters, `kernel1` and `kernel2`, that define the `conv1` and `conv2` layers:

```
class TfLeNetMultiTrialModel(Model):

    def __init__(self, kernel1, kernel2):
        super().__init__()
        self.conv1 = create_conv(kernel1, filter = 16)
        self.pool1 = MaxPool2D(pool_size = 2)
        self.conv2 = create_conv(kernel2, filter = 32)
        self.pool2 = MaxPool2D(pool_size = 2)
        self.flatten = Flatten()
        self.fc1 = Dense(128, 'relu')
        self.fc2 = Dense(10, 'softmax')

    def call(self, x, **kwargs):
        x = self.conv1(x)
        x = self.pool1(x)
        x = self.conv2(x)
        x = self.pool2(x)
        x = self.flatten(x)
        x = self.fc1(x)
        return self.fc2(x)
```

We import necessary modules:

Listing 5-2. **PyTorch** implementation. Model Space. ch5/naive_one_shot_nas/pt/pt_lenet_multi_model.py

```
import torch
import torch.nn as nn
import torch.nn.functional as F
from ch5.naive_one_shot_nas.pt.pt_ops import create_conv
```

The following model below accepts two parameters, kernel1 and kernel2, that define the conv1 and conv2 layers:

```
class PtLeNetMultiTrialModel(nn.Module):

    def __init__(self, kernel1, kernel2):
        super(PtLeNetMultiTrialModel, self).__init__()

        self.conv1 = create_conv(kernel1, in_channels = 1, out_channels = 16)
        self.conv2 = create_conv(kernel2, in_channels = 16, out_channels = 32)
        self.flat = nn.Flatten()
        self.fc1 = nn.Linear(1568, 128)
        self.fc2 = nn.Linear(128, 10)

    def forward(self, x):
        x = torch.relu(self.conv1(x))
        x = F.max_pool2d(x, 2, 2)
        x = torch.relu(self.conv2(x))
        x = F.max_pool2d(x, 2, 2)
        x = self.flat(x)
        x = torch.relu(self.fc1(x))
        x = self.fc2(x)
        return F.log_softmax(x, dim = 1)
```

And now, let's execute Multi-trial NAS iterating through the various kernel_size parameters (kernel1: [1, 3, 5], kernel2: [1, 3, 5]) using the script in Listing 5-3 (TensorFlow implementation) and Listing 5-4 (PyTorch implementation).

We import necessary modules:

Listing 5-3. **TensorFlow** implementation. Multi-trial NAS. ch5/naive_one_shot_nas/tf/ms_search.py

```
from ch5.naive_one_shot_nas.tf.tf_lenet_multi_model import
TfLeNetMultiTrialModel
from ch5.naive_one_shot_nas.tf.tf_train import train, test
```

Defining the search space:

```
kernel1_choices = [1, 3, 5]
kernel2_choices = [1, 3, 5]

results = {}
```

Performing Multi-trial search:

```
for k1 in kernel1_choices:
    for k2 in kernel2_choices:
        # Trial
        model = TfLeNetMultiTrialModel(k1, k2)
        train(model)
        accuracy = test(model)
        results[(k1, k2)] = accuracy
```

Displaying results:

```
print('=======')
print('Results:')
for k, v in results.items():
    print(f'Conv1 {k[0]}x{k[0]}, Conv2: {k[1]}x{k[1]} : {v}')
```

We import necessary modules:

Listing 5-4. **PyTorch** implementation. Multi-trial NAS. ch5/naive_one_shot_nas/pt/ms_search.py

```
from ch5.naive_one_shot_nas.pt.pt_lenet_multi_model import
PtLeNetMultiTrialModel
from ch5.naive_one_shot_nas.pt.pt_train import train_model, test_model
```

Defining the search space:

```
kernel1_choices = [1, 3, 5]
kernel2_choices = [1, 3, 5]

results = {}
```

Performing Multi-trial search:

```
for k1 in kernel1_choices:
    for k2 in kernel2_choices:
        # Trial
        model = PtLeNetMultiTrialModel(k1, k2)
        train_model(model)
        accuracy = test_model(model)
        results[(k1, k2)] = accuracy
```

Defining the search space:

```
print('=======')
print('Results:')
for k, v in results.items():
    print(f'Conv1 {k[0]}x{k[0]}, Conv2: {k[1]}x{k[1]} : {v}')
```

The results of the Multi-trial NAS we performed are listed in Table 5-1.

Table 5-1. *Multi-trial NAS results*

Trial	Conv1	Conv2	Accuracy
1	Conv1×1	Conv1×1	0.9446
2	Conv1×1	Conv3×3	0.9849
3	Conv1×1	Conv5×5	0.9864
4	Conv3×3	Conv3×3	0.9851
5	Conv3×3	Conv3×3	0.9881
6	Conv3×3	Conv5×5	0.9909
7	Conv5×5	Conv1×1	0.9872
8	Conv5×5	Conv3×3	0.9901
9	**Conv5×5**	**Conv5×5**	**0.9917**

According to Table 5-1, the best candidate is (`Conv5×5, Conv5×5`). Well, we tried every neural design candidate and found the most suitable one for the MNIST problem. Of course, the Multi-trial NAS we implemented earlier is quite simple, but that's how all Multi-trial approaches act in general.

But for now, let's try to get the same result using the One-shot NAS approach! First, we create the Supernet model depicted in Figure 5-2 (in Listing 5-5 [TensorFlow implementation] and Listing 5-6 [PyTorch implementation]).

We import necessary modules:

Listing 5-5. **TensorFlow** implementation. Supernet. ch5/naive_one_shot_nas/tf/tf_lenet_supernet.py

```
from tensorflow.keras import Model
from tensorflow.keras.layers import Dense, Flatten, MaxPool2D
from ch5.naive_one_shot_nas.tf.tf_ops import create_conv
```

`TfLeNetNaiveSupernet` implements Supernet depicted in Figure 5-2:

```
class TfLeNetNaiveSupernet(Model):

    def __init__(self):
        super().__init__()
```

We define each candidate for `conv1` and `conv2` layers:

```
        self.conv1_1 = create_conv(1, 16)
        self.conv1_3 = create_conv(3, 16)
        self.conv1_5 = create_conv(5, 16)

        self.conv2_1 = create_conv(1, 32)
        self.conv2_3 = create_conv(3, 32)
        self.conv2_5 = create_conv(5, 32)
```

Next are the other Supernet layers:

```
        self.pool1 = MaxPool2D(pool_size = 2)
        self.pool2 = MaxPool2D(pool_size = 2)
        self.flatten = Flatten()
        self.fc1 = Dense(128, 'relu')
        self.fc2 = Dense(10, 'softmax')
```

call method accepts mask parameter, which activates candidate layer in a sum merge operation. The mask parameter is not passed in the training mode, and all candidates are summed:

```
x = 1×conv1_1(x) + 1×conv1_3(x) + 1×conv1_5(x)
x = 1×conv2_1(x) + 1×conv2_3(x) + 1×conv2_5(x).
```

But in evaluation mode, we pass mask parameter and activate only particular layers:

```
x = 0×conv1_1(x) + 1×conv1_3(x) + 0×conv1_5(x)
x = 0×conv2_1(x) + 0×conv2_3(x) + 1×conv2_5(x):
    def call(self, x, mask = None):

        # Sum all in training mode
        if mask is None:
            mask = [[1, 1, 1], [1, 1, 1]]

        x = mask[0][0] * self.conv1_1(x) +\
            mask[0][1] * self.conv1_3(x) +\
            mask[0][2] * self.conv1_5(x)
        x = self.pool1(x)

        x = mask[1][0] * self.conv2_1(x) +\
            mask[1][1] * self.conv2_3(x) +\
            mask[1][2] * self.conv2_5(x)
        x = self.pool2(x)

        x = self.flatten(x)
        x = self.fc1(x)
        return self.fc2(x)
```

We import necessary modules:

Listing 5-6. **PyTorch** implementation. Supernet. ch5/naive_one_shot_nas/pt/pt_lenet_supernet.py

```
import torch
import torch.nn as nn
import torch.nn.functional as F
from ch5.naive_one_shot_nas.pt.pt_ops import create_conv
```

PtLeNetNaiveSupernet implements Supernet depicted in Figure 5-2:

```
class PtLeNetNaiveSupernet (nn.Module):

    def __init__(self):
        super(PtLeNetNaiveSupernet, self).__init__()
```

We define each candidate for conv1 and conv2 layers:

```
        self.conv1_1 = create_conv(1, 1, 16)
        self.conv1_3 = create_conv(3, 1, 16)
        self.conv1_5 = create_conv(5, 1, 16)

        self.conv2_1 = create_conv(1, 16, 32)
        self.conv2_3 = create_conv(3, 16, 32)
        self.conv2_5 = create_conv(5, 16, 32)
```

Next are the other Supernet layers:

```
        self.flat = nn.Flatten()
        self.fc1 = nn.Linear(1568, 128)
        self.fc2 = nn.Linear(128, 10)
```

forward method accepts mask parameter, which activates candidate layer in a sum merge operation. The mask parameter is not passed in the training mode, and all candidates are summed:

```
x = 1×conv1_1(x) + 1×conv1_3(x) + 1×conv1_5(x)
x = 1×conv2_1(x) + 1×conv2_3(x) + 1×conv2_5(x).
```

But in evaluation mode, we pass mask parameter and activate only particular layers:

```
x = 0×conv1_1(x) + 1×conv1_3(x) + 0×conv1_5(x)
x = 0×conv2_1(x) + 0×conv2_3(x) + 1×conv2_5(x):
    def forward(self, x, mask = None):

        # Sum all in training mode
        if mask is None:
            mask = [[1, 1, 1], [1, 1, 1]]

        x = mask[0][0] * self.conv1_1(x) +\
            mask[0][1] * self.conv1_3(x) +\
            mask[0][2] * self.conv1_5(x)
```

```
        x = torch.relu(x)
        x = F.max_pool2d(x, 2, 2)

        x = mask[1][0] * self.conv2_1(x) +\
            mask[1][1] * self.conv2_3(x) +\
            mask[1][2] * self.conv2_5(x)
        x = torch.relu(x)
        x = F.max_pool2d(x, 2, 2)

        x = self.flat(x)
        x = torch.relu(self.fc1(x))
        x = self.fc2(x)

        return F.log_softmax(x, dim = 1)
```

Next, we train the Supernet and evaluate different candidate layer combinations in Listing 5-7 (TensorFlow implementation) and Listing 5-8 (PyTorch implementation).

We import necessary modules:

Listing 5-7. **TensorFlow** implementation. One-shot NAS. ch5/naive_one_shot_nas/tf/os_search.py

```
import tensorflow as tf
from sklearn.metrics import accuracy_score
from ch5.datasets import mnist_dataset
from ch5.naive_one_shot_nas.tf.tf_lenet_supernet import
TfLeNetNaiveSupernet
from ch5.naive_one_shot_nas.tf.tf_train import train
```

Initializing Supernet:

```
model = TfLeNetNaiveSupernet()
```

Training Supernet:

```
train(model)
```

Loading test dataset:

```
_, (x, y) = mnist_dataset()
```

Evaluating Supernet activating each candidate:

```
kernel1_choices = [1, 3, 5]
kernel2_choices = [1, 3, 5]

results = {}
for m1 in range(0, len(kernel1_choices)):
    for m2 in range(0, len(kernel2_choices)):
        # activation mask
        mask = [[0, 0, 0], [0, 0, 0]]
        # activating conv1 and conv2 layers
        mask[0][m1] = 1
        mask[1][m2] = 1

        # calculating accuracy
        output = model(x, mask = mask)
        predict = tf.argmax(output, axis = 1)
        accuracy = round(accuracy_score(predict, y), 4)
        results[(kernel1_choices[m1], kernel2_choices[m2])] = accuracy
```

Displaying results:

```
print('=======')
print('Results:')
for k, v in results.items():
    print(f'Conv1 {k[0]}x{k[0]}, Conv2: {k[1]}x{k[1]} : {v}')
```

We import necessary modules:

Listing 5-8. **PyTorch** implementation. One-shot NAS. ch5/naive_one_shot_nas/pt/os_search.py

```
import torch
from sklearn.metrics import accuracy_score
from ch5.datasets import mnist_dataset
from ch5.naive_one_shot_nas.pt.pt_lenet_supernet import
PtLeNetNaiveSupernet
from ch5.naive_one_shot_nas.pt.pt_train import train_model
```

Initializing Supernet:

```
model = PtLeNetNaiveSupernet()
```

Training Supernet:

```
train_model(model)
```

Loading test dataset:

```
_, (x, y) = mnist_dataset()
x = torch.from_numpy(x).float()
y = torch.from_numpy(y).long()
x = torch.permute(x, (0, 3, 1, 2))
```

Evaluating Supernet activating each candidate:

```
model.eval()
kernel1_choices = [1, 3, 5]
kernel2_choices = [1, 3, 5]
results = {}
for m1 in range(0, len(kernel1_choices)):
    for m2 in range(0, len(kernel2_choices)):
        # activation mask
        mask = [[0, 0, 0], [0, 0, 0]]
        # activating conv1 and conv2 layers
        mask[0][m1] = 1
        mask[1][m2] = 1

        # calculating accuracy
        output = model(x, mask)
        predict = output.argmax(dim = 1, keepdim = True)
        accuracy = round(accuracy_score(predict, y), 4)
        results[(kernel1_choices[m1], kernel2_choices[m2])] = accuracy
```

Displaying results:

```
print('=======')
print('Results:')
for k, v in results.items():
    print(f'Conv1 {k[0]}x{k[0]}, Conv2: {k[1]}x{k[1]} : {v}')
```

The results of the One-shot NAS are listed in Table 5-2.

Table 5-2. *One-shot NAS results*

Trial	Conv1	Conv2	Accuracy
1	Conv1×1	Conv1×1	0.2343
2	Conv1×1	Conv3×3	0.1789
3	Conv1×1	Conv5×5	0.2127
4	Conv3×3	Conv3×3	0.2755
5	Conv3×3	Conv3×3	0.8515
6	Conv3×3	Conv5×5	0.8786
7	Conv5×5	Conv1×1	0.3486
8	Conv5×5	Conv3×3	0.8882
9	**Conv5×5**	**Conv5×5**	**0.9001**

The best neural architecture found by One-shot NAS is (`Conv 5×5, Conv 5×5`), which is exactly the same as the result of Multi-trial NAS. Incredible, isn't it? We found the same result in a much shorter time!

Note The results presented in Table 5-2 are only needed to rank various combinations of architectures to pick the best one. These results do not characterize the accuracy of the corresponding combination. One-shot models are typically only used to rank architectures in the Model Space. The best-performing architectures are retrained from scratch after the search is completed.

Intuitively, the One-shot NAS approach can be illustrated as follows: *during Supernet training, the best performing candidate layers play much more significant roles in the Data Flow Graph of the Supernet neural network.* This fact allows finding these candidates by deactivating others during the evaluation process. Figure 5-4 demonstrates this concept.

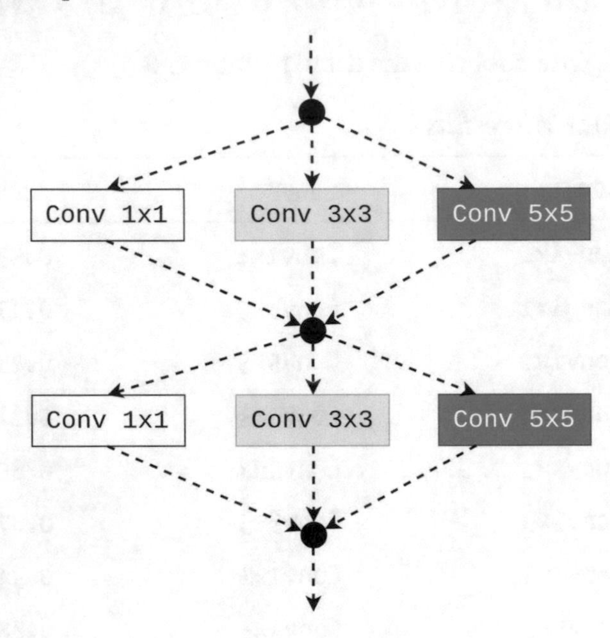

***Figure 5-4.** Different layer importance in the Supernet*

Since we have an intuition about the One-shot NAS approach, we can start implementing it using the NNI framework.

Supernet Architecture

As we saw in the previous section, one of the main concepts in One-shot NAS is the *Supernet*. Supernet is a single neural network that contains all the various neural network architectures from the defined Model Space. Supernet is trained once according to the One-shot NAS technique, and then the optimal subnet is selected. In Multi-trial NAS, each Data Flow Graph is tried separately. But One-shot NAS creates a single Supernet based on all possible Data Flow Graphs in Model Space.

NNI creates Model Space for One-shot NAS using `LayerChoice` and `InputChoice` operations. `LayerChoice` candidates form a special block in the Supernet. Each `LayerChoice` candidate transforms input tensor, and then their output tensors are reduced according to a particular One-shot NAS algorithm. The reduce operation can be `sum`, `mean`, or any other operation that merges tensors. `InputChoice` candidate tensors are reduced in Supernet in the same way as `LayerChoice` candidates. Figure 5-5 demonstrates Model Space constructed with `LayerChoice` and `InputChoice` operations.

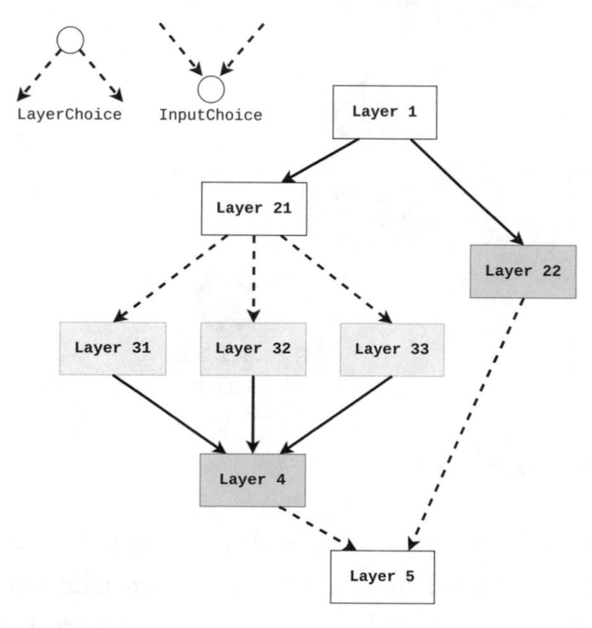

Figure 5-5. *Model Space*

Model Space depicted in Figure 5-5 generates Supernet with reduce operations. Figure 5-6 demonstrates Supernet with `sum` as reduce operations.

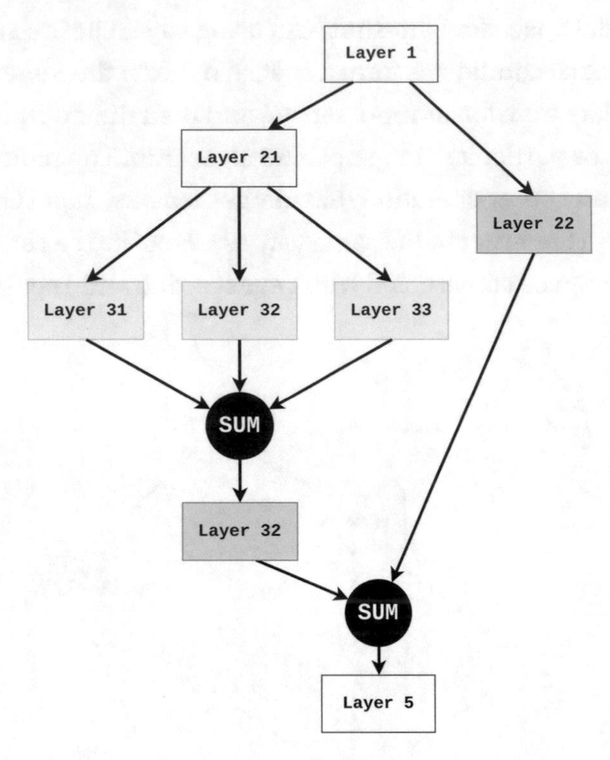

Figure 5-6. *One-shot NAS Supernet*

Supernet is trained according to a specific One-shot NAS algorithm. After the training, each subnet is evaluated, and the subnetwork with the best accuracy forms the target neural architecture. Figure 5-7 demonstrates the evaluation of a specific subnet.

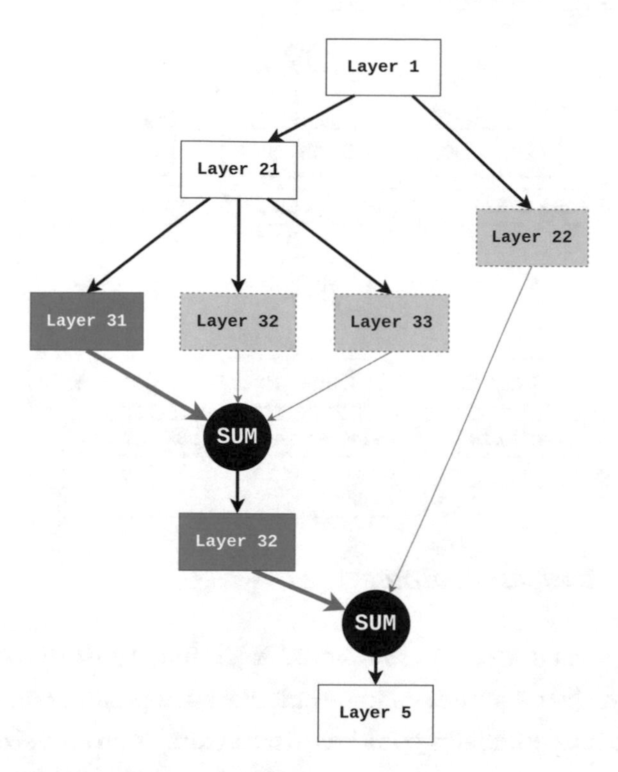

Figure 5-7. *Subnet Evaluation*

One-shot NAS sets a strict restriction on the `LayerChoice` and `InputChoice` candidates. Each of the candidates must return a tensor of the same size. Otherwise, it will be impossible to implement the reduce operation. In the previous section, candidates for `conv1` layer returned 16×28×28 tensors, and candidates for `conv2` layer returned 32×14×14 tensors. Figure 5-8 demonstrates this fact.

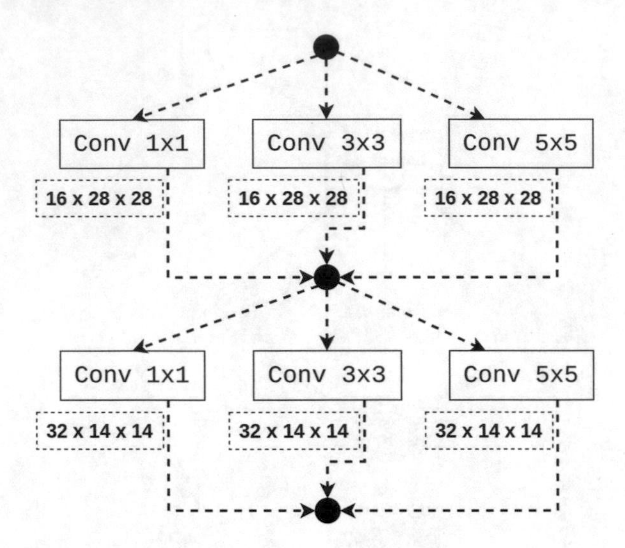

Figure 5-8. *Same size tensor outputs*

This restriction does not exist in Multi-trial NAS since TensorFlow and PyTorch frameworks allowed layer parameters to be calculated depending on the input tensor; therefore, `LayerChoice` candidates could return tensors of various sizes. In the case of One-shot NAS, we must be sure that the candidates return tensors of the same size. Otherwise, the NAS algorithm fails with an error.

Let's create our first Model Space for One-shot NAS. It will be a "*Hello World*" model, which we will use to test One-shot NAS algorithms in the next section. We will define Model Space for One-shot search using LeNet architecture variations depicted in Figure 5-9.

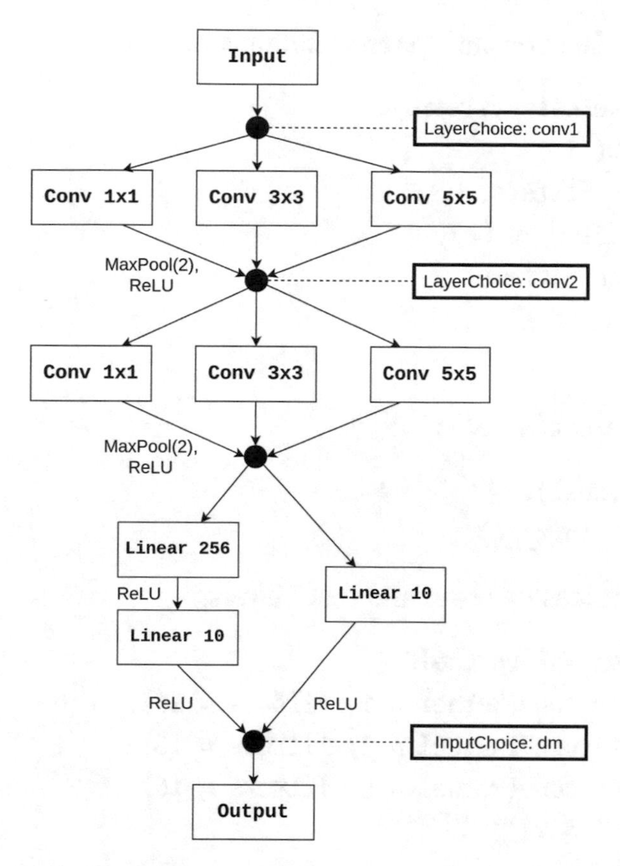

Figure 5-9. *LeNet One-shot Model Space*

NNI implementation for TensorFlow of the Model Space depicted in Figure 5-9 is provided in Listing 5-9.

LayerChoice and InputChoice methods are implemented in nni.nas.tensorflow.mutables package:

Listing 5-9. **TensorFlow** implementation. One-shot LeNet NAS. ch5/model/lenet/tf_lenet.py

```
from nni.nas.tensorflow.mutables import InputChoice, LayerChoice
```

Importing `tensorflow.keras` modules:

```
from tensorflow.keras import Model
from tensorflow.keras.layers import Conv2D, Dense, Flatten, MaxPool2D
```

Defining helper function creating convolution layer:

```
def create_conv(kernel, filters):
    return Conv2D(
        filters = filters,
        kernel_size = kernel,
        activation = 'relu',
        padding = 'same'
    )

class TfLeNetSupernet(Model):

    def __init__(self):
        super().__init__()
```

Setting LayerChoices for conv1 and conv2 layers:

```
        self.conv1 = LayerChoice([
            create_conv(kernel = 1, filters = 16),  # 0
            create_conv(kernel = 3, filters = 16),  # 1
            create_conv(kernel = 5, filters = 16)  # 2
        ], key = 'conv1')

        self.conv2 = LayerChoice([
            create_conv(kernel = 1, filters = 32),  # 0
            create_conv(kernel = 3, filters = 32),  # 1
            create_conv(kernel = 5, filters = 32)  # 2
        ], key = 'conv2')

        self.pool = MaxPool2D(2)
        self.flat = Flatten()
```

Setting InputChoice for linear layers:

```
        self.dm = InputChoice(n_candidates = 2, n_chosen = 1, key = 'dm')

        self.fc11 = Dense(256, activation = 'relu')
        self.fc12 = Dense(10, activation = 'softmax')
        self.fc2 = Dense(10, activation = 'softmax')
```

Defining call method:

```
def call(self, x):
    x = self.conv1(x)
    x = self.pool(x)
    x = self.conv2(x)
    x = self.pool(x)
    x = self.flat(x)

    # branch 1
    x1 = self.fc12(self.fc11(x))
    # branch 2
    x2 = self.fc2(x)

    # Choosing one of the branches
    x = self.dm([
        x1,  # 0
        x2  # 1
    ])

    return x
```

NNI implementation for PyTorch of the Model Space depicted in Figure 5-9 is provided in Listing 5-10.

LayerChoice and InputChoice methods are implemented in nni.retiarii.nn.pytorch package:

Listing 5-10. **PyTorch** implementation. One-shot LeNet NAS. ch5/model/lenet/pt_lenet.py

```
from nni.retiarii.nn.pytorch import LayerChoice, InputChoice
```

Importing other modules:

```
from typing import OrderedDict
import torch.nn as nn
import torch.nn.functional as F
```

Defining helper function creating convolution layer:

```
def create_conv(kernel, in_ch, out_ch):
    return nn.Conv2d(
        in_channels = in_ch,
        out_channels = out_ch,
        kernel_size = kernel,
        padding = int((kernel - 1) / 2)
    )

class PtLeNetSupernet(nn.Module):

    def __init__(self, input_ts = 32):
        super(PtLeNetSupernet, self).__init__()
```

Setting LayerChoices for conv1 and conv2 layers:

```
        self.conv1 = LayerChoice(OrderedDict(
            [
                ('conv1x1->16', create_conv(1, 1, 16)),  # 0
                ('conv3x3->16', create_conv(3, 1, 16)),  # 1
                ('conv5x5->16', create_conv(5, 1, 16)),  # 2
            ]
        ), label = 'conv1')

        self.conv2 = LayerChoice(OrderedDict(
            [
                ('conv1x1->32', create_conv(1, 16, 32)),  # 0
                ('conv3x3->32', create_conv(3, 16, 32)),  # 1
                ('conv5x5->32', create_conv(5, 16, 32)),  # 2
            ]
        ), label = 'conv2')

        self.act = nn.ReLU()
        self.flat = nn.Flatten()
```

Setting InputChoice for linear layers:

```
self.dm = InputChoice(n_candidates = 2, n_chosen = 1, label = 'dm')

self.fc11 = nn.Linear(input_ts * 8 * 8, 256)
self.fc12 = nn.Linear(256, 10)
self.fc2 = nn.Linear(input_ts * 8 * 8, 10)
```

Defining forward method:

```
def forward(self, x):
    x = self.act(self.conv1(x))
    x = F.max_pool2d(x, 2, 2)
    # x.shape = (16, 16, 16)
    x = self.act(self.conv2(x))
    x = F.max_pool2d(x, 2, 2)
    # x.shape = (32, 8, 8)

    x = self.flat(x)

    # branch 1
    x1 = self.act(self.fc11(x))
    x1 = self.act(self.fc12(x1))

    # branch 2
    x2 = self.act(self.fc2(x))

    # Choosing one of the branches
    x = self.dm([
        x1,  # 0
        x2  # 1
    ])

    return F.log_softmax(x, dim = 1)
```

Since we defined One-shot Model Space using NNI, we can move on to implementing advanced One-shot algorithms.

One-Shot Algorithms

At the moment, One-shot NAS is a young area and is developing rapidly. Many algorithms that implement the One-shot concept are being invented at the moment. This section will study two of the most popular One-shot algorithms: Efficient Neural Architecture Search (ENAS) and Differentiable Architecture Search (DARTS).

Efficient Neural Architecture Search (ENAS)

Efficient Neural Architecture Search (ENAS) is a fast and inexpensive approach for automatic model design. In ENAS, a controller discovers neural network architectures by searching for an optimal subnet within a large Supernet. The controller is trained with a policy gradient to select a subnet that maximizes the expected reward on a validation set. Meanwhile, the model corresponding to the selected subnet is trained to minimize the loss function. Sharing parameters among child models allows ENAS to deliver strong empirical performances. It uses much fewer GPU hours than classical Multi-trial NAS approaches.

The original paper "Efficient Neural Architecture Search via Parameter Sharing" (*https://arxiv.org/pdf/1802.03268.pdf*) has a lot of formulas and can be too complicated. Let's study the idea of this approach in a more practical way.

One of the key concepts in ENAS is Reinforcement Learning Controller or RL Controller or simply Controller. RL Controller contains a neural network θ that learns how to extract the most efficient subnets from the Supernet. The main task of the RL Controller is to find the optimal subnet in the Supernet, and it is trained according to the Reinforcement Learning algorithm. The Controller creates a subnet by defining a binary mask, where 1 activates layer and 0 disables it. Figure 5-10 shows how RL Controller selects subnet from the Supernet using binary mask.

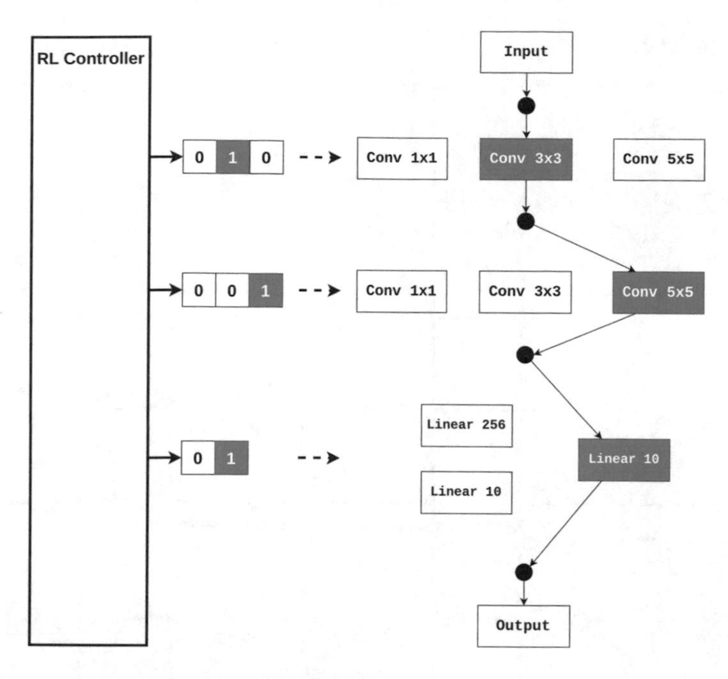

Figure 5-10. *RL Controller subnet selection*

RL Controller picks a subnet and runs one or several training epochs. The core of the ENAS approach is the weight-sharing technique. The same layers in different subnets share the same weights. When Controller picks a subnet, it is not being trained from scratch; layers share weights that have already been trained. Weight-sharing concept is demonstrated in Figure 5-11.

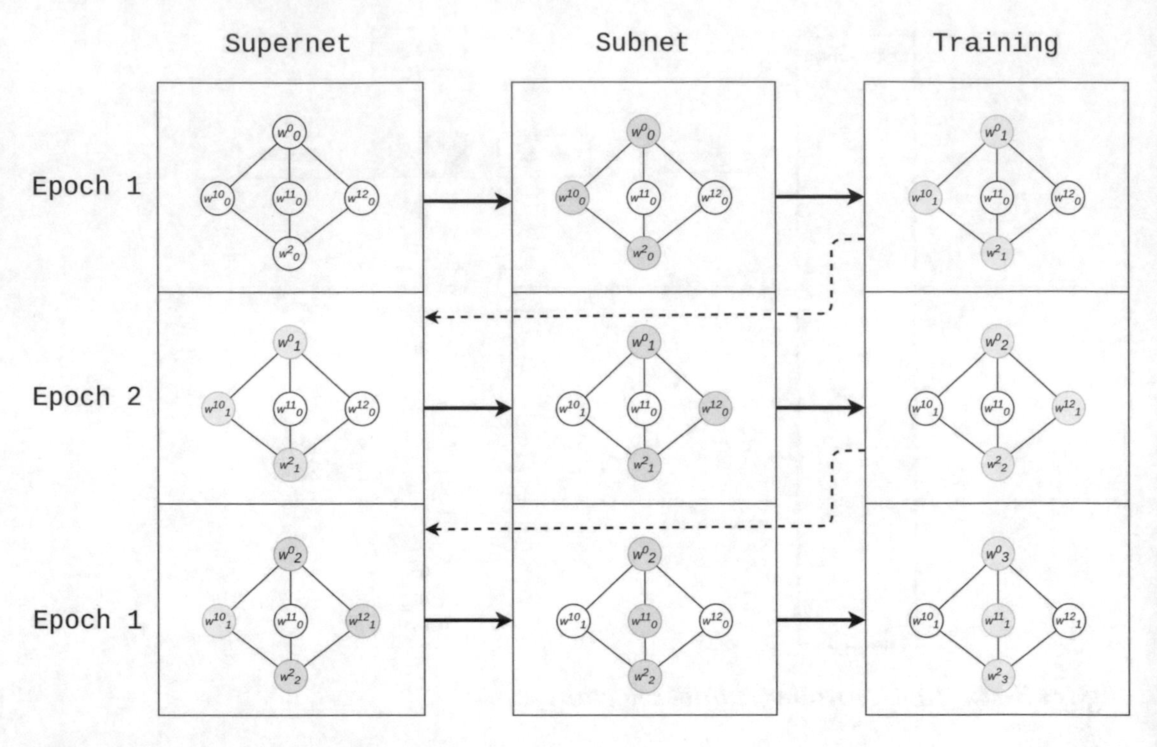

Figure 5-11. *Weight sharing*

The weight-sharing approach allows the Controller to find the best architecture with small iterations and without retraining a new subnet from scratch each time. ENAS algorithm can be demonstrated using the following pseudo-code:

We initialize a Supernet: `S`

And load train and validation datasets: `train_ds, val_ds`

Next, the algorithm initializes ENAS Controller (Controller(S, θ)), where θ denotes Controller weights that help choose optimal subnet from S Supernet:

```
Ctrl = Controller(S, θ)
```

Main training loop:

```
for epoch in epochs:
```

1. Supernet training loop:

   ```
   for batch in batches(train_ds):
   ```

i. Controller implements stochastic policy, which means that it operates with probabilities. The most promising subnet has the highest probability to be chosen:

```
s ← Ctrl.sample() #picks pseudo-random subnet
```

ii. Subnet network is trained once (i.e., only one training epoch) on a training batch using weight-sharing technique (Figure 5-11):

```
train_once(subnet, batch)
```

2. Controller training loop. In this loop, Controller learns how to find a subnet that gives the highest reward, that is, accuracy on validation dataset:

```
for batch in batches(val_ds):
```

i. Controller chooses subnet and calculates its accuracy on validation dataset batch:

```
subnet ← Ctrl.sample() #picks pseudo random subnet
reward = test(subnet, batch)
```

ii. Controller collects experience about subnet performance:

```
Ctrl.add_experience(reward)
```

3. Controller updates its weights θ according to new experience:

```
Ctrl.self_update_with_new_experience()
```

```
# end of main training loop
```

After training, Controller returns the best subnet:

```
Ctrl.best()
```

In the algorithm described earlier, the Controller trains various subnets on the training dataset and then tests them on the validation dataset. By repeating this process many times, the Controller understands which architectures show the best accuracy and gradually reduces the exploration process to a limited number of architectures. At the end of the training process, the Controller converges to one or several of the best architectures.

Let's see how ENAS works in practice:

1. Initially, the Controller has no assumptions about subnets, and it chooses random subnets for single training (subnet is trained for one training epoch only) using the weight-sharing technique as depicted in Figure 5-11.

2. After, Controller generates subnets based on θ weights and collects accuracy on the validation dataset.

3. Based on experience gained in step 2, Controller updates θ values.

Figure 5-12 demonstrates steps 1–3 we described earlier.

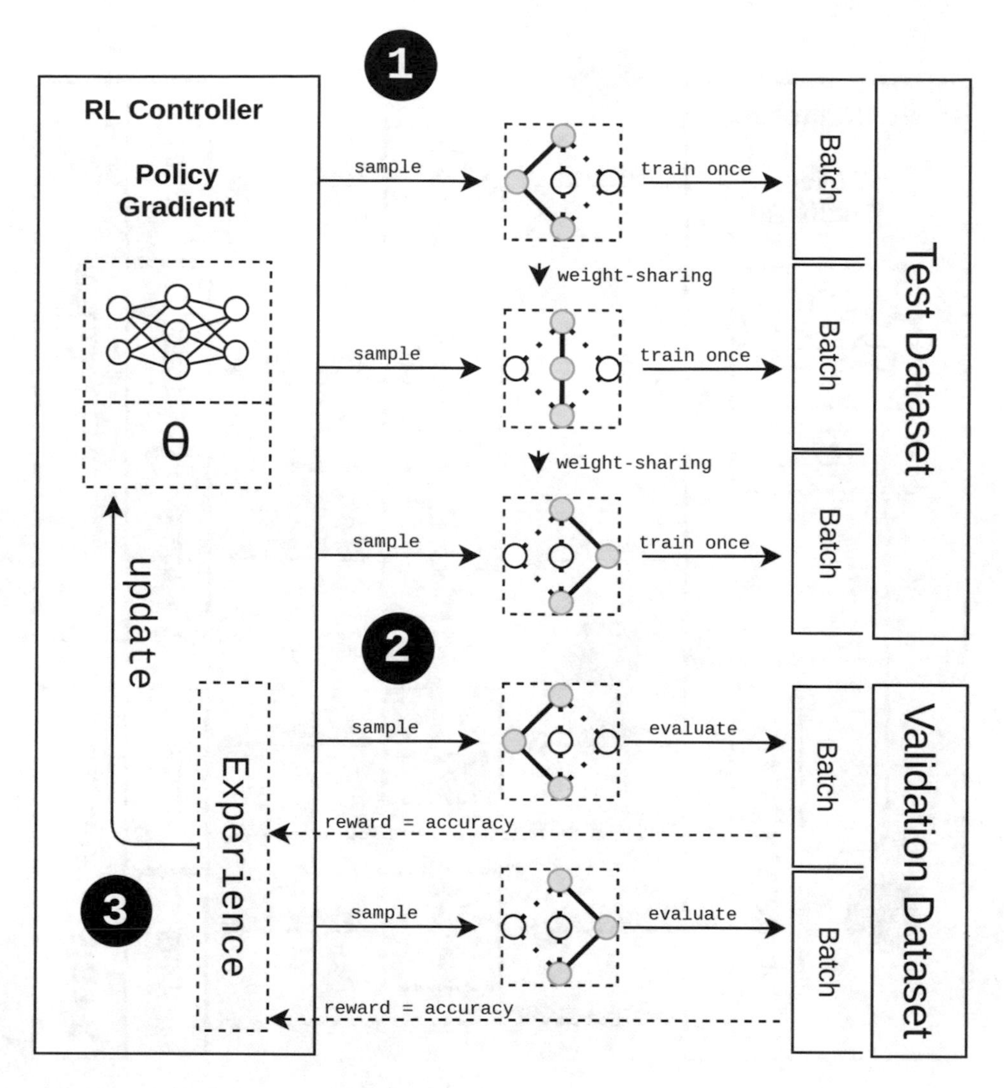

Figure 5-12. *ENAS in action. Initial epoch*

Next, Controller chooses the most promising subnets to train using the weight-sharing technique and tests their accuracy on the validation dataset, as shown in Figure 5-13.

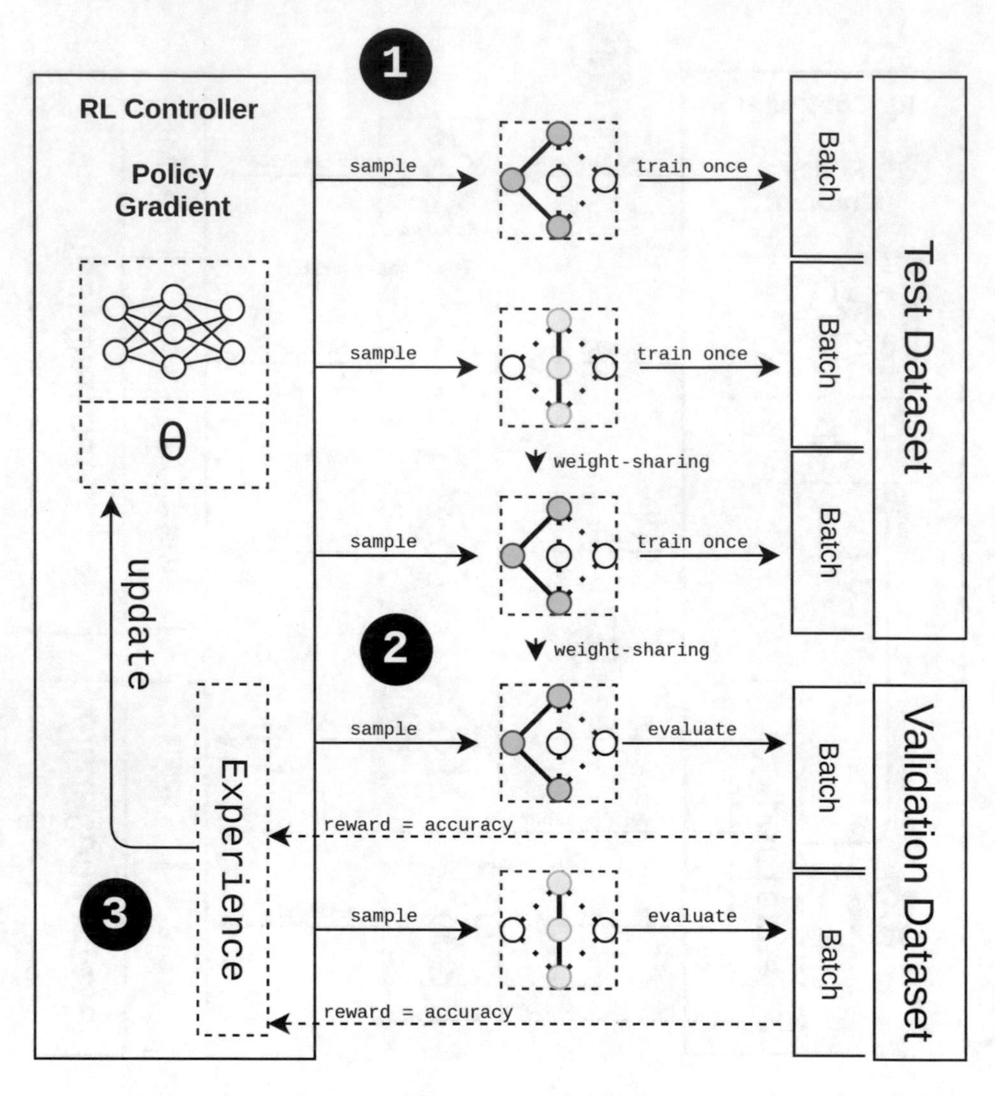

***Figure 5-13.** ENAS in action. Middle training*

After training Supernet layers separately and updating θ experience, Controller converges to some subnet that it considers the best. Figure 5-14 illustrates it.

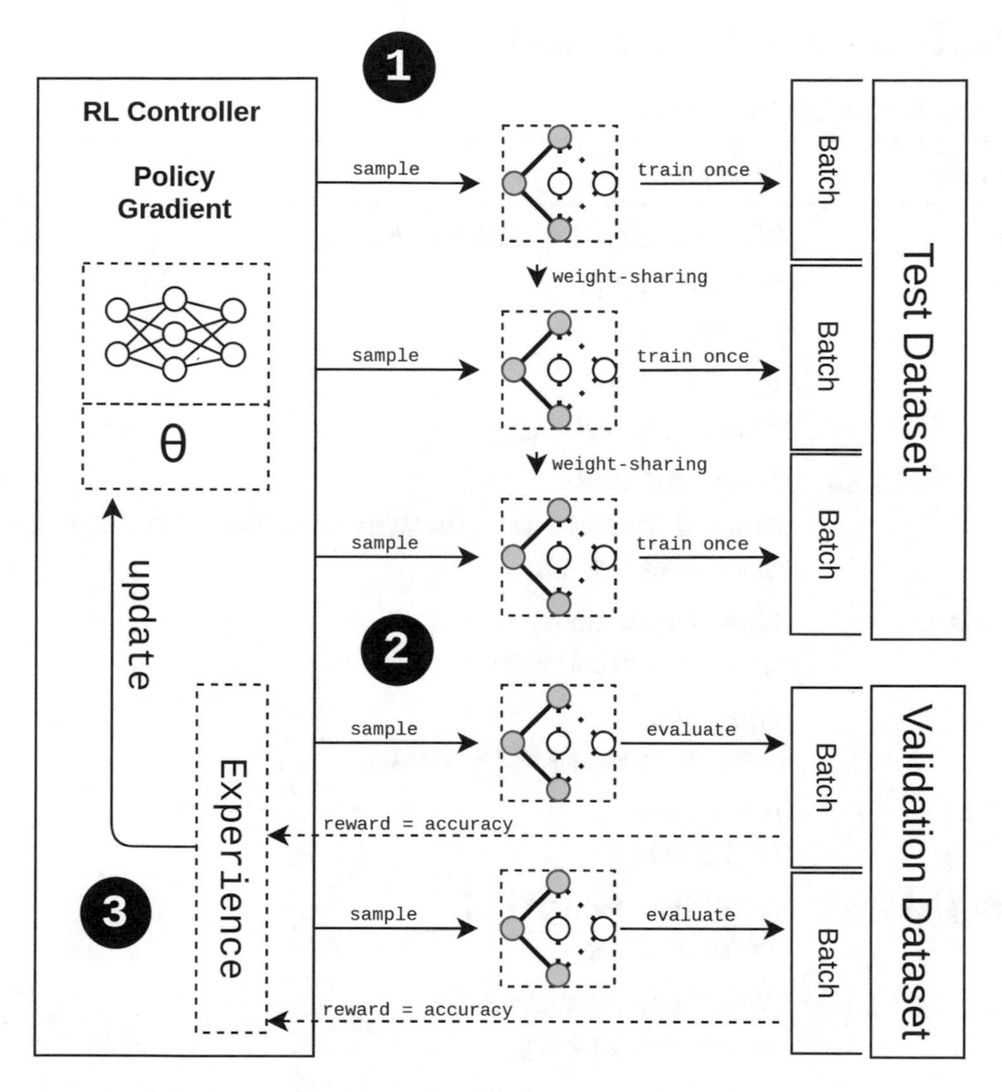

***Figure 5-14.** ENAS in action. End training*

One-shot ENAS algorithm is close to RL Strategy from Multi-trial NAS, but with one significant difference. ENAS does not make a complete subnet training cycle but uses weight sharing and incremental one-step training. This difference dramatically speeds up the process of finding the best architecture.

NNI implements ENAS using the following classes:

- **PyTorch**: `nni.retiarii.oneshot.pytorch.enas.EnasTrainer`
- **TensorFlow**: `nni.algorithms.nas.tensorflow.enas.EnasTrainer`

Table 5-3 shows EnasTrainer parameters.

Table 5-3. *EnasTrainer parameters*

Parameter	Description
model	PyTorch or TensorFlow model to be trained
loss	Type: callable Loss function
metrics	Type: callable Measures model accuracy
reward_function	Type: callable Is used by ENAS Controller to calculate reward. Usually, reward_function returns model accuracy
optimizer	Type: Optimizer Optimizer for model training
num_epochs	Type: int Number of epochs planned for training
dataset	Type: Dataset Training dataset
batch_size	Type: int, Default: 64 Training batch size
workers	Type: int, Default: 4 Workers for data loading
log_frequency	Type: int, Default: None Logging step count
grad_clip	Type: float, Default: 5.0 Gradient clipping. Set to 0 to disable
entropy_weight	Type: float, Default: 0.0001 Sample entropy loss weight
skip_weight	Type: float, Default: 0.8 Skip penalty loss weight

(*continued*)

Table 5-3. (*continued*)

Parameter	Description
baseline_decay	Type: float, Default: 0.999 Baseline decay rate. New baseline is calculated as baseline_decay * baseline_old + reward * (1 - baseline_decay
ctrl_lr	Type: float, Default: 0.00035 Controller learning rate
ctrl_steps_aggregate	Type: int, Default: 20 Number of steps that will be aggregated into one mini-batch for Controller

Now let's move on to the practical application of the ENAS algorithm for the LeNet Model Space we defined in the previous section.

TensorFlow ENAS Implementation

Listing 5-11 demonstrates ENAS application using TensorFlow LeNet Model Space.

Importing modules:

Listing 5-11. ENAS TensorFlow. ch5/enas/enas_tf_search.py

```
from tensorflow.keras.losses import Reduction,
SparseCategoricalCrossentropy
from tensorflow.keras.optimizers import Adam
import ch5.datasets as datasets
from nni.algorithms.nas.tensorflow import enas
from ch5.model.lenet.tf_lenet import TfLeNetSupernet
from ch5.tf_utils import accuracy, reward_accuracy, get_best_model
```

Initializing LeNetSupernet:

```
model = TfLeNetSupernet()
```

Loading datasets:

```
dataset_train, dataset_valid = datasets.mnist_dataset()
```

Defining loss function:

```
loss = SparseCategoricalCrossentropy(
    from_logits = True,
    reduction = Reduction.NONE
)
```

Defining optimizer:

```
optimizer = Adam()
```

ENAS training params:

```
num_epochs = 10
batch_size = 256
```

Initializing `EnasTrainer`:

```
trainer = enas.EnasTrainer(
    model,
    loss = loss,
    metrics = accuracy,
    reward_function = reward_accuracy,
    optimizer = optimizer,
    batch_size = batch_size,
    num_epochs = num_epochs,
    dataset_train = dataset_train,
    dataset_valid = dataset_valid,
    log_frequency = 10,
    child_steps = 10,
    mutator_steps = 30
)
```

Launching One-shot search:

```
trainer.train()
```

Returning best subnet:

```
best = get_best_model(trainer.mutator)
print(best)
```

Listing 5-11 returns the following best model as the result of ENAS algorithm:

- `conv1: 1 (Conv3×3)`
- `conv2: 2 (Conv5×5)`
- `dm: 0 (Linear256→Linear10)`

PyTorch ENAS Implementation

Listing 5-12 demonstrates ENAS application using PyTorch LeNet Model Space.

Importing modules:

Listing 5-12. ENAS PyTorch. ch5/enas/enas_pt_search.py

```
import torch.nn as nn
from nni.retiarii.oneshot.pytorch.enas import EnasTrainer
from torch.optim.sgd import SGD
import ch5.datasets as datasets
from ch5.model.lenet.pt_lenet import PtLeNetSupernet
from ch5.pt_utils import accuracy, reward_accuracy
```

Initializing LeNetSupernet:

```
model = PtLeNetSupernet()
```

Loading datasets:

```
dataset_train, dataset_valid = datasets.get_dataset("mnist")
```

Defining loss function:

```
criterion = nn.CrossEntropyLoss()
```

Defining optimizer:

```
optimizer = SGD(
    model.parameters(), 0.05,
    momentum = 0.9, weight_decay = 1.0E-4
)
```

ENAS training params:

```
batch_size = 256
log_frequency = 50
num_epochs = 10
ctrl_kwargs = {"tanh_constant": 1.1}
```

Initializing EnasTrainer:

```
trainer = EnasTrainer(
    model,
    loss = criterion,
    metrics = accuracy,
    reward_function = reward_accuracy,
    optimizer = optimizer,
    batch_size = batch_size,
    num_epochs = num_epochs,
    dataset = dataset_train,
    log_frequency = log_frequency,
    ctrl_kwargs = ctrl_kwargs,
    ctrl_steps_aggregate = 20
)
```

Launching One-shot search:

```
trainer.fit()
```

Returning best subnet:

```
best_model = trainer.export()
print(best_model)
```

Listing 5-12 returns the following best model as the result of ENAS algorithm:

- `conv1: 1 (Conv3×3)`
- `conv2: 1 (Conv3×3)`
- `dm: 0 (Linear256→Linear10)`

ENAS is one of the first One-shot NAS algorithms that made the community rethink the whole approach to Neural Architecture Search. But ENAS may seem complicated to an inexperienced reader due to the complex internal algorithm and nontrivial tuning. In the next section, we'll study a more elegant One-shot NAS technique.

Differentiable Architecture Search (DARTS)

From calculus, we know that one of the most efficient ways to find the maxima of a continuous differentiable surface is to use derivatives and gradient-based methods. Neural networks use gradient descent based on the principle of computing derivatives. It would also be very convenient to reduce NAS to a differentiation problem, and the DARTS algorithm does that. DARTS algorithm was proposed in the original paper "DARTS: Differentiable Architecture Search" (`https://arxiv.org/abs/1806.09055`).

Using binary masks, ENAS algorithm chooses subnets from the Supernet, but this approach is discrete. ENAS Controller jumps from one subnet to another, trying to discover the best one. DARTS algorithm makes the search space continuous; it relaxes the categorical choice of a particular operation to a softmax over all possible operations:

$$o'(x) = \sum_i \frac{\exp(\alpha_i) o_i(x)}{\sum_j \exp(\alpha_j)}$$

This means that the DARTS algorithm creates a Supernet derived from Model Space with sum reducing, and each choice operation is followed with α_i parameter that specifies the weight of the operation. This makes the $\{\alpha\}$ parameter set trainable as Supernet weights. At the end of Supernet training, the choices with the highest α values are chosen as the optimal subnet operations. Figure 5-15 illustrates this concept.

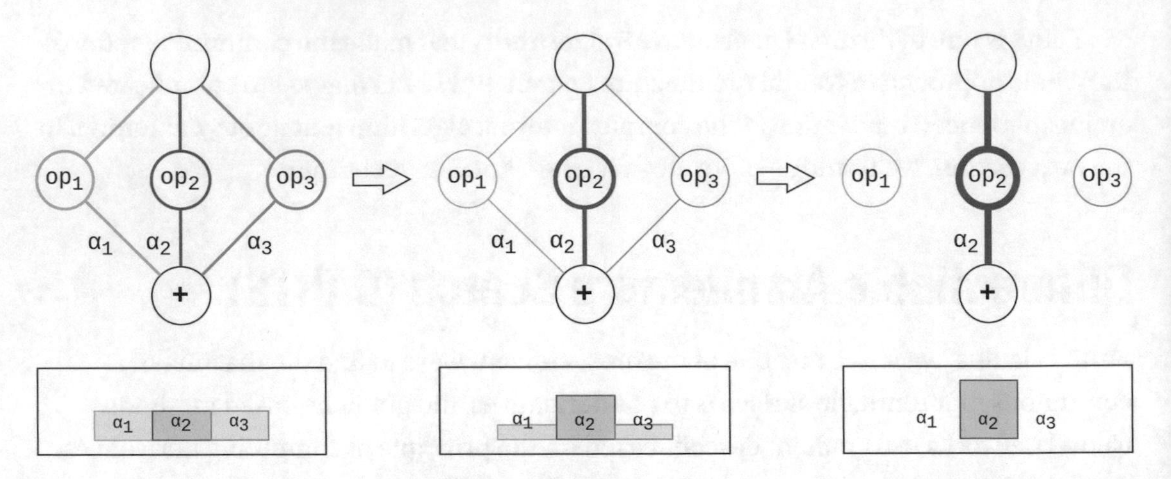

***Figure 5-15.** DARTS operation relaxation*

During Supernet training using the DARTS algorithm, inefficient choices tend to be zeroed, and the search converges to a single architecture, which is the search result. Figure 5-16 visualizes the DARTS algorithm applied to the LeNetSupermodel. It gradually relaxes inefficient layers showing the best architecture.

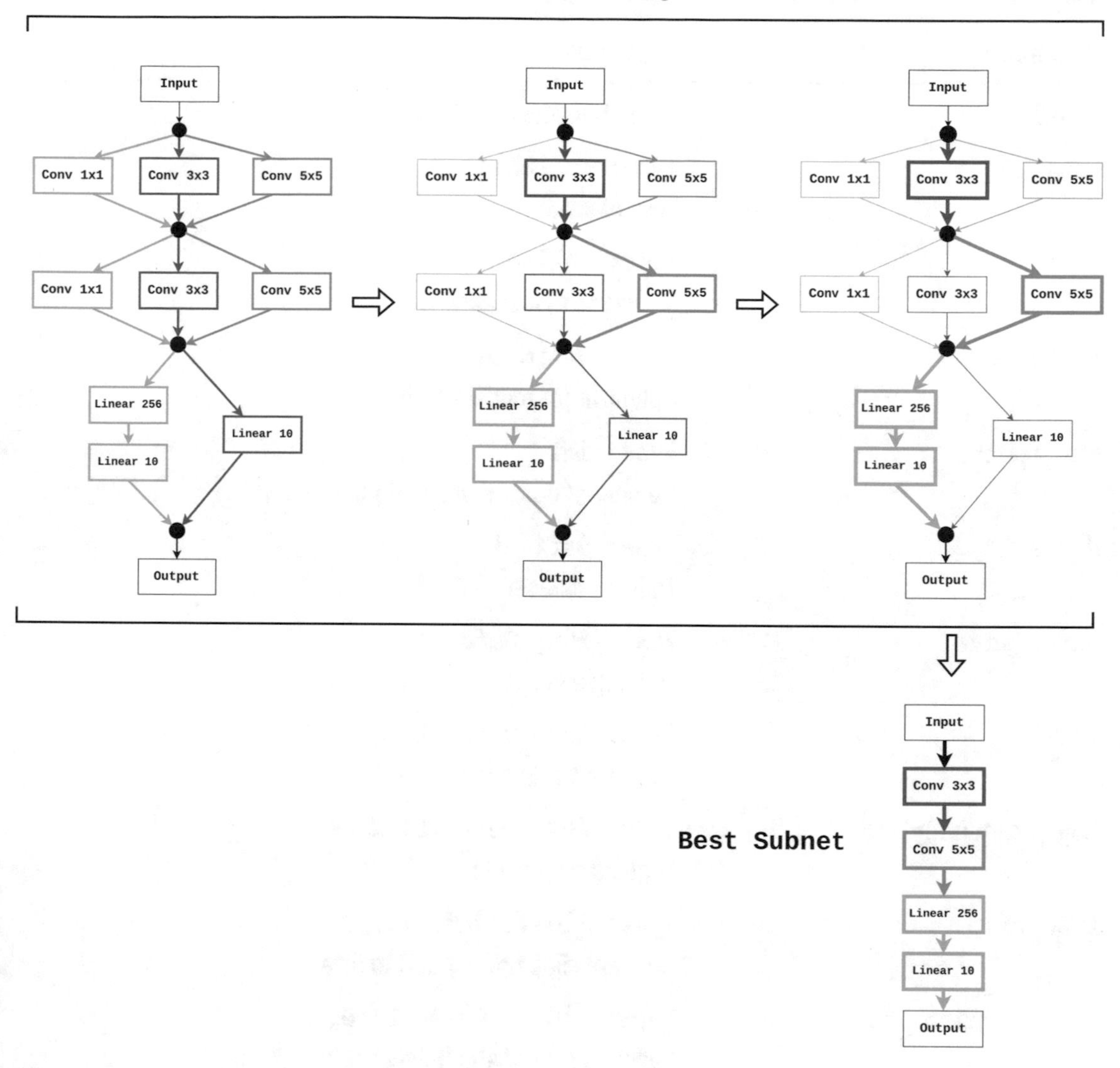

***Figure 5-16.** DARTS in action*

NNI 2.7 implements DARTS only for PyTorch framework using the following class: `nni.retiarii.oneshot.pytorch.DartsTrainer`.

Table 5-4 shows `DartsTrainer` parameters.

Table 5-4. *DartsTrainer parameters*

Parameter	Description
`model`	PyTorch model to be trained
`loss`	`Type: callable` Loss function
`metrics`	`Type: callable` Measures model accuracy
`optimizer`	`Type: Optimizer` Optimizer for model training
`num_epochs`	`Type: int` Number of epochs planned for training
`dataset`	`Type: Dataset` Training dataset
`batch_size`	`Type: int, Default: 64` Training batch size
`workers`	`Type: int, Default: 4` Workers for data loading
`log_frequency`	`Type: int, Default: None` Logging step count
`grad_clip`	`Type: float, Default: 5.0` Gradient clipping. Set to 0 to disable
`arc_learning_rate`	`Type: float, Default: 0.0001` Learning rate of architecture parameters

Listing 5-13 implements DARTS algorithm using NNI.

Importing modules:

Listing 5-13. DARTS PyTorch. ch5/darts/darts_pt_search.py

```
import torch
import torch.nn as nn
import ch5.datasets as datasets
from nni.retiarii.oneshot.pytorch import DartsTrainer
```

```
from ch5.model.lenet.pt_lenet import PtLeNetSupernet
from ch5.pt_utils import accuracy
```

Initializing LeNetSupernet:

```
model = PtLeNetSupernet()
```

Loading datasets:

```
dataset_train, dataset_valid = datasets.get_dataset("mnist")
```

Defining loss function:

```
criterion = nn.CrossEntropyLoss()
```

Defining optimizer:

```
optim = torch.optim.SGD(
    model.parameters(), 0.025,
    momentum = 0.9, weight_decay = 3.0E-4
)
```

ENAS training params:

```
num_epochs = 10
batch_size = 256
metrics = accuracy
```

Initializing DartsTrainer:

```
trainer = DartsTrainer(
    model = model,
    loss = criterion,
    metrics = metrics,
    optimizer = optim,
    num_epochs = num_epochs,
    dataset = dataset_train,
    batch_size = batch_size,
    log_frequency = 10,
    unrolled = False
)
```

Launching One-shot search:

```
trainer.fit()
```

Returning best subnet:

```
best_architecture = trainer.export()
print('Best architecture:', best_architecture)
```

Listing 5-13 returns the following best model as the result of DARTS algorithm:

- `conv1: conv5x5->16`
- `conv2: conv5x5->32`
- `dm: 1 (Linear10)`

DARTS is a clear and straightforward One-shot algorithm. It has an intuitive logic and is easy to tune. But DARTS requires more memory than ENAS because DARTS trains the whole Supernet, while ENAS trains the various subnets only. Anyway, DARTS is a good choice to implement Neural Architecture Search.

GeneralSupernet Solving CIFAR-10

We considered the application of One-shot algorithms on the simplest LeNet Model Space solving the trivial MNIST problem. These examples are good as entry points, but they don't show the power of the One-shot NAS approach. In this section, we will examine a more complex Model Space for solving the CIFAR-10 problem.

Usually, One-shot NAS deals with cell-designed Supernets. As the name suggests, a cell-designed Supernet consists of various cells. Each **cell** accepts a different number of inputs and creates a computational graph using a deep learning block operation inside. Each **block operation** is a `LayerChoice` of deep learning layers. Let's build a cell-designed Supernet called **GeneralSupernet** to solve the CIFAR-10 problem.

In GeneralSupernet, we define block operation as a `LayerChoice` from

- `SepConvBranch(3)`
- `NonSepConvBranch(3)`
- `SepConvBranch(5)`
- `NonSepConvBranch(3)`

- `AvgPoolBranch`
- `MaxPoolBranch`

The implementations of these layers are not provided here, but the reader can get details in the following source code files:

- **TensorFlow**: ch5/model/general/tf_ops.py
- **PyTorch**: ch5/model/general/pt_ops.py

Figure 5-17 depicts block operation space.

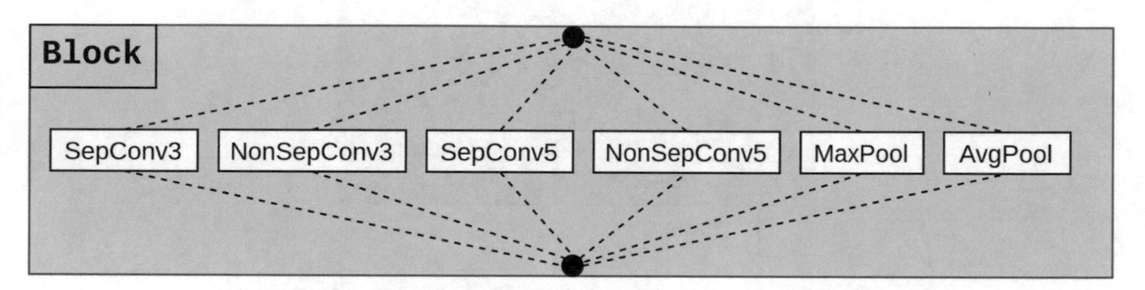

***Figure 5-17.** GeneralSupernet block operation*

Each *n*th cell accepts one required input which is transformed by block operation and *n* additional inputs. Additional inputs are not required and can be zeroed in different subnets. The normalized sum of block operation output and additional inputs forms cell output. Figure 5-18 demonstrates examples of cell spaces.

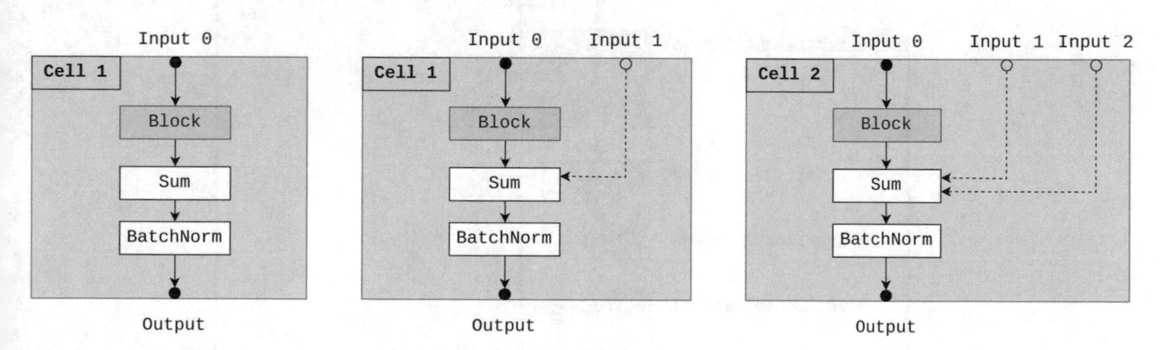

***Figure 5-18.** GeneralSupernet cell*

The sequence of cells forms the GeneralSupernet, and the output of each cell can be the input of the subsequent cell. After every three cells, a `FactorizedReduced` layer is inserted. In this section, we will use a GeneralSupernet with six cells. Figure 5-19 depicts GeneralSupernet architecture.

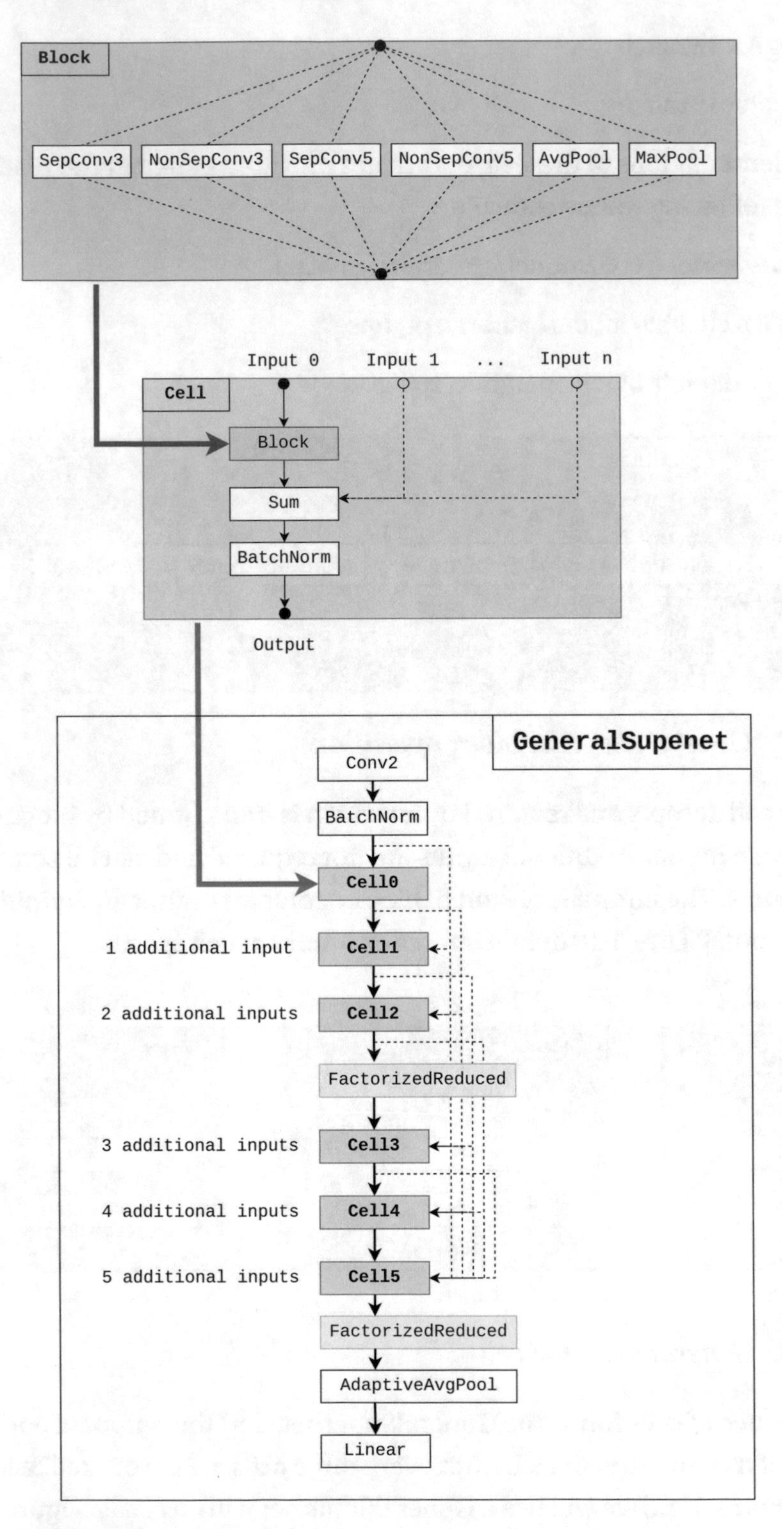

Figure 5-19. *GeneralSupernet*

Let's calculate how many subnets GeneralSupernet has: $(6) \times (6\times2) \times (6\times2^2) \times (6\times2^3) \times (6\times2^4) \times (6\times2^5) = 6^6 \times 2^{1+2+3+4+5} \sim 1{,}500{,}000{,}000$. Of course, it is not possible to efficiently explore this Model Space using a Multi-trial NAS approach. It is also possible to use the GeneralSupernet with 9, 12, or 24 cells. In this case, the number of subnets will become enormous.

There are a lot of predefined Supernets for One-shot NAS that are aimed at solving a specific class of problems. And GeneralSupernet we defined earlier is one of the simplest. Let's implement GeneralSupernet and run One-shot NAS on the CIFAR-10 dataset.

Training GeneralSupernet Using TensorFlow and ENAS

Let's implement GeneralSupernet and find the best architecture using ENAS algorithm. Listing 5-14 defines GeneralSupernet using TensorFlow.

Importing modules:

Listing 5-14. GeneralSupernet. TensorFlow. ch5/model/general/tf_general.py

```
from tensorflow.keras import Model, Sequential
from tensorflow.keras.layers import BatchNormalization, Conv2D, Dense,
GlobalAveragePooling2D
from nni.nas.tensorflow.mutables import InputChoice, LayerChoice,
MutableScope
from ch5.model.general.tf_ops import build_conv, build_separable_conv,
build_avg_pool, build_max_pool, FactorizedReduce
```

Defining `Cell` depicted in Figure 5-18:

```
class Cell(MutableScope):

    def __init__(self, cell_ord, input_num, filters):
        super().__init__(f'cell_{cell_ord}')
```

Setting LayerChoice for block operation depicted in Figure 5-17:

```
self.block_op = LayerChoice([
    build_conv(filters, 3, 'conv3'),
    build_separable_conv(filters, 3, 'sepconv3'),
    build_conv(filters, 5, 'conv5'),
    build_separable_conv(filters, 5, 'sepconv5'),
    build_avg_pool(filters, 'avgpool'),
    build_max_pool(filters, 'maxpool'),
], key = f'op_{cell_ord}')
```

Setting InputChoice for additional Cell inputs:

```
if input_num > 0:
    self.connections = InputChoice(
        n_candidates = input_num,
        n_chosen = None,
        key = f'con_{cell_ord}'
    )
else:
    self.connections = None
```

Last cell layer – BatchNormalization:

```
self.batch_norm = BatchNormalization(trainable = False)
```

Defining call method:

```
def call(self, inputs):
```

Main input is processed by block_op:

```
out = self.block_op(inputs[-1])
```

Additional inputs are selected by self.connections and summed:

```
if self.connections is not None:
    connection = self.connections(inputs[:-1])
    if connection is not None:
        out += connection

return self.batch_norm(out)
```

Defining GeneralSupernet depicted in Figure 5-19:

```
class GeneralSupernet(Model):

    def __init__(
            self,
            num_cells = 6,
            filters = 24,
            num_classes = 10
    ):
        super().__init__()
        self.num_cells = num_cells

        self.stem = Sequential([
            Conv2D(filters, kernel_size = 3, padding = 'same', use_bias
            = False),
            BatchNormalization()
        ])
```

Setting the positions for pool layers (FactorizedReduce):

```
        # num_cells = 6 -> pool_layers_idx = [3, 6]
        self.pool_layers_idx = [
            cell_id
            for cell_id in range(1, num_cells + 1) if cell_id % 3 == 0
        ]
```

Initializing cells and pool_layers lists:

```
        self.cells = []
        self.pool_layers = []

        for cell_ord in range(num_cells):
            if cell_ord in self.pool_layers_idx:
                pool_layer = FactorizedReduce(filters)
                self.pool_layers.append(pool_layer)
            cell = Cell(cell_ord, cell_ord, filters)
            self.cells.append(cell)
```

Defining final layers:

```
        self.gap = GlobalAveragePooling2D()
        self.dense = Dense(num_classes)
```

Next, we define call method:

```
    def call(self, x):
        cur = self.stem(x)
        prev_outputs = [cur]

        for cell_id, cell in enumerate(self.cells):
```

Passing Cell outputs through FactorizedReduce pooling layer:

```
            if cell_id in self.pool_layers_idx:
                # Number of Pool Layer
                # 0, 1, 2, ....
                pool_ord = self.pool_layers_idx.index(cell_id)
                pool = self.pool_layers[pool_ord]
                prev_outputs = [pool(tensor) for tensor in prev_outputs]
                cur = prev_outputs[-1]

            cur = cell(prev_outputs)
            prev_outputs.append(cur)

        cur = self.gap(cur)
        logits = self.dense(cur)
        return logits
```

Since we defined GeneralSupernet, we can launch ENAS using the following script. Importing modules:

Listing 5-15. GeneralSupernet ENAS. TensorFlow. ch5/cifar10/enas_tf.py

```
from tensorflow.keras.losses import Reduction, SparseCategoricalCrossentropy
from tensorflow.keras.optimizers import SGD
import ch5.datasets as datasets
from nni.algorithms.nas.tensorflow import enas
from ch5.model.general.tf_general import GeneralSupernet
from ch5.tf_utils import accuracy, reward_accuracy, get_best_model
```

Initializing GeneralSupernet:

```
model = GeneralSupernet()
```

Loading datasets:

```
dataset_train, dataset_valid = datasets.cifar10_dataset()
```

Declaring loss function:

```
loss = SparseCategoricalCrossentropy(
    from_logits = True,
    reduction = Reduction.NONE
)
```

Declaring optimizer:

```
optimizer = SGD(learning_rate = 0.05, momentum = 0.9)
```

Setting ENAS trainer parameters:

```
metrics = accuracy
reward_function = reward_accuracy
batch_size = 256
num_epochs = 100
```

Initializing `EnasTrainer`:

```
trainer = enas.EnasTrainer(
    model,
    loss = loss,
    metrics = metrics,
    reward_function = reward_function,
    optimizer = optimizer,
    batch_size = batch_size,
    num_epochs = num_epochs,
    dataset_train = dataset_train,
    dataset_valid = dataset_valid
)
```

Launching training:

```
trainer.train()
```

Displaying results:

```
best = get_best_model(trainer.mutator)
print(best)
```

Note Duration ~ 6 hours on Intel Core i7 with CUDA (GeForce GTX 1050)

After the search was completed, the following report was returned:

- `op_layer_0: 3 SepConvBranch(5)`
- `op_layer_1: 3 SepConvBranch(5)`
- `op_layer_2: 1 SepConvBranch(3)`
- `op_layer_3: 4 NonSepConvBranch(3)`
- `op_layer_4: 1 SepConvBranch(3)`
- `op_layer_5: 1 SepConvBranch(3)`
- `con_layer_1: 0 Additional input from Cell0`
- `con_layer_2: 0 Additional input from Cell0`
- `con_layer_3: None No additional inputs`
- `con_layer_4: [0, 2, 3] Additional inputs from: Cell0, Cell2, Cell3`
- `con_layer_5: 3 Additional input from Cell3`

Figure 5-20 visualizes the result returned by ENAS.

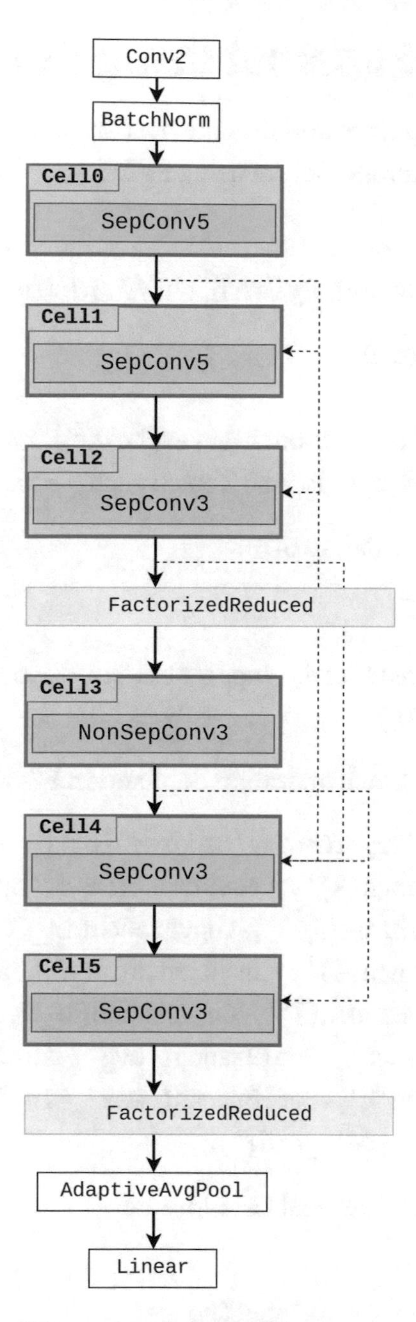

***Figure 5-20.** ENAS GeneralSupernet best architecture*

As see in Figure 5-20, the best architecture does not use pooling operations, `AvgPoolBranch` and `MaxPoolBranch`, and this makes sense because GeneralSupernet has built-in `FactorizedReduced` layers.

Training GeneralSupernet Using PyTorch and DARTS

Let's implement GeneralSupernet and find the best architecture using DARTS algorithm. First, we need to define GeneralSupernet using PyTorch.

Importing modules:

Listing 5-16. GeneralSupernet. PyTorch. ch5/model/general/pt_general.py

```
from typing import OrderedDict
import torch.nn as nn
from nni.retiarii.nn.pytorch import LayerChoice, InputChoice
from ch5.model.general.pt_ops import ConvBranch, PoolBranch, FactorizedReduce
```

Defining Cell depicted in Figure 5-18:

```
class Cell(nn.Module):

    def __init__(self, cell_ord, input_num, in_f, out_f):
        super().__init__()
```

Setting LayerChoice for block operation depicted in Figure 5-17:

```
        self.block_op = LayerChoice(OrderedDict([
            ('SepConvBranch(3)', ConvBranch(in_f, out_f, 3, 1, 1, False)),
            ('NonSepConvBranch(3)', ConvBranch(in_f, out_f, 3, 1, 1, True)),
            ('SepConvBranch(5)', ConvBranch(in_f, out_f, 5, 1, 2, False)),
            ('NonSepConvBranch(3)', ConvBranch(in_f, out_f, 5, 1, 2, True)),
            ('AvgPoolBranch', PoolBranch('avg', in_f, out_f, 3, 1, 1)),
            ('MaxPoolBranch', PoolBranch('max', in_f, out_f, 3, 1, 1))
        ]), label = f'op_{cell_ord}')
```

Setting InputChoice for additional Cell inputs:

```
        if input_num > 0:
            self.connections = InputChoice(
                n_candidates = input_num, n_chosen = None,
                label = f'con_{cell_ord}'
            )
        else:
            self.connections = None
```

Last cell layer – BatchNormalization:

```
self.batch_norm = nn.BatchNorm2d(out_f, affine = False)
```

Defining forward method:

```
def forward(self, inputs):
```

Main input is processed by block_op:

```
out = self.block_op(inputs[-1])
```

Additional inputs are selected by self.connections and summed:

```
if self.connections is not None:
    connection = self.connections(inputs[:-1])
    if connection is not None:
        out = out + connection

return self.batch_norm(out)
```

Defining GeneralSupernet depicted in Figure 5-19:

```
class GeneralSupernet(nn.Module):

    def __init__(
            self,
            num_cells = 6,
            out_f = 24,
            in_channels = 3,
            num_classes = 10
    ):
        super().__init__()
        self.num_cells = num_cells

        # Stem layer
        self.stem = nn.Sequential(
            nn.Conv2d(in_channels, out_f, 3, 1, 1, bias = False),
            nn.BatchNorm2d(out_f)
        )
```

Setting the positions for pool layers (FactorizedReduce):

```
self.pool_layers_idx = [
    cell_id
    for cell_id in range(1, num_cells + 1) if cell_id % 3 == 0
]
```

Initializing cells and pool_layers lists:

```
self.cells = nn.ModuleList()
self.pool_layers = nn.ModuleList()

# Initializing Cells and Pool Layers
for cell_ord in range(num_cells):
    if cell_ord in self.pool_layers_idx:
        pool_layer = FactorizedReduce(out_f, out_f)
        self.pool_layers.append(pool_layer)
    cell = Cell(cell_ord, cell_ord, out_f, out_f)
    self.cells.append(cell)
```

Defining final layers:

```
self.gap = nn.AdaptiveAvgPool2d(1)
self.dense = nn.Linear(out_f, num_classes)
```

Next, we define forward method:

```
def forward(self, x):
    bs = x.size(0)
    cur = self.stem(x)

    # Constructing Calculation Graph
    cells = [cur]
    for cell_id in range(self.num_cells):
        cur = self.cells[cell_id](cells)
        cells.append(cur)
```

Passing `Cell` outputs through `FactorizedReduce` pooling layer:

```
            # If pool layer is added
            if cell_id in self.pool_layers_idx:
                # Number of Pool Layer
                # 0, 1, 2, ...
                pool_ord = self.pool_layers_idx.index(cell_id)
                # Adding Pool Layer to all input cells
                for i, cell in enumerate(cells):
                    cells[i] = self.pool_layers[pool_ord](cell)
                cur = cells[-1]

        cur = self.gap(cur).view(bs, -1)
        logits = self.dense(cur)
        return logits
```

Since we defined GeneralSupernet, we can launch DARTS using Listing 5-17.

Importing modules:

Listing 5-17. GeneralSupernet DARTS. PyTorch. ch5/cifar10/darts_pt.py

```
import torch
import torch.nn as nn
import ch5.datasets as datasets
from nni.retiarii.oneshot.pytorch import DartsTrainer
from ch5.model.general.pt_general import GeneralSupernet
from ch5.pt_utils import accuracy
```

Initializing GeneralSupernet:

```
model = GeneralSupernet()
```

Loading datasets:

```
dataset_train, dataset_valid = datasets.get_dataset("cifar10")
```

Declaring loss function:

```
criterion = nn.CrossEntropyLoss()
```

Declaring optimizer:

```
optim = torch.optim.SGD(
    model.parameters(), 0.025,
    momentum = 0.9, weight_decay = 3.0E-4
)
```

Setting DARTS trainer parameters:

```
num_epochs = 100
batch_size = 128
accuracy_metrics = accuracy
```

Initializing DartsTrainer:

```
trainer = DartsTrainer(
    model = model,
    loss = criterion,
    metrics = accuracy_metrics,
    optimizer = optim,
    num_epochs = num_epochs,
    dataset = dataset_train,
    batch_size = batch_size,
    log_frequency = 10,
    unrolled = False
)
```

Launching training:

```
trainer.fit()
```

Displaying results:

```
best_architecture = trainer.export()
print('Best architecture:', best_architecture)
```

Note Duration ~ 4 hours on Intel Core i7 with CUDA (GeForce GTX 1050)

After the search was completed, the following report was returned:

- `op_layer_0: SepConvBranch(3)`
- `op_layer_1: SepConvBranch(5)`
- `op_layer_2: SepConvBranch(5)`
- `op_layer_3: SepConvBranch(5)`
- `op_layer_4: SepConvBranch(5)`
- `op_layer_5: MaxPoolBranch`
- `con_layer_1: 0 Additional input from Cell0`
- `con_layer_2: 0 Additional input from Cell0`
- `con_layer_3: 0 Additional input from Cell0`
- `con_layer_4: 2 Additional input from Cell2`
- `con_layer_5: 4 Additional input from Cell4`

Figure 5-21 visualizes the result returned by DARTS.

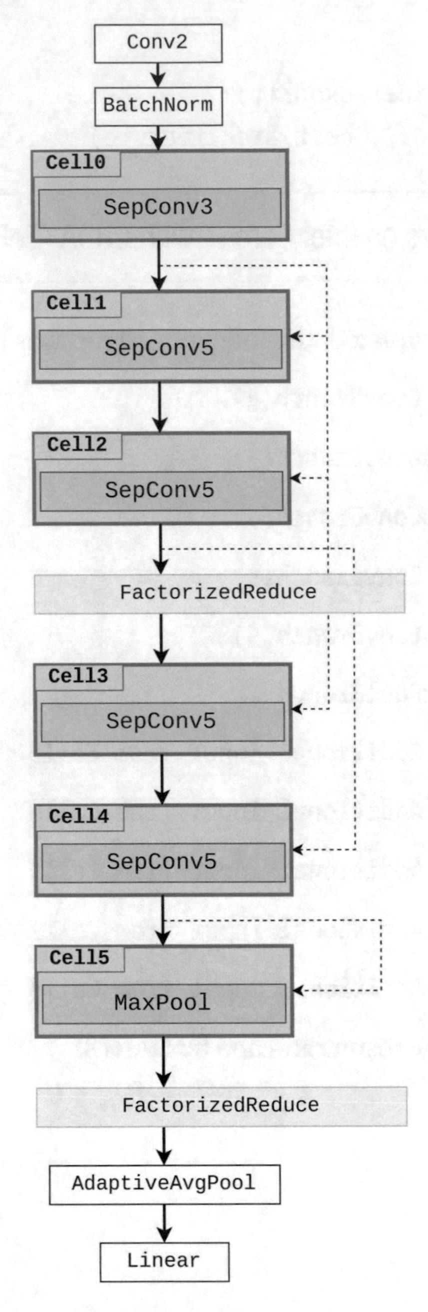

Figure 5-21. *DARTS GeneralSupernet best architecture*

The architectures obtained using ENAS (Figure 5-20) and DARTS (Figure-21) are similar. They tend to use the `SepConvBranch(5)` operation and share `Cell0` output. ENAS best architecture and DARTS best architecture achieve 91.2% and 92.8% accuracy, respectively. But we can further improve the accuracy if we increase the number of cells (`num_cells`) in the GeneralSupernet. This will make the search longer, but it will result in a more accurate target architecture. The beautiful thing is that we can use the same GeneralSupernet and One-shot algorithm for any pattern recognition problem. This gives us a universal approach to solving typical deep learning problems. Absolutely One-shot NAS is one of the most significant achievements of automated deep learning.

HPO vs. Multi-trial NAS vs. One-Shot NAS

And so, at the moment, we have three different approaches to optimize and construct deep learning models: HPO, Multi-trial NAS, and One-shot NAS. And a fair question may arise: which method is better to choose? But this question does not have a clear answer. Each approach is better suited for a specific task.

- **HPO** deals with black-box optimization and is suitable for selecting optimization algorithms, training batch size, and tuning the predesigned model. HPO results can be visualized and easily analyzed. HPO can be a good starting point to dive into a completely new problem or a good finishing point when the final tuning is performed.
- **Multi-trial NAS** is only concerned with finding the optimal architecture. Despite its durability, Multi-trial NAS is more accurate than One-shot NAS. Multi-trial NAS results are much easier to interpret because this approach provides metrics for each tried architecture.
- **One-shot NAS** is well suited for finding complex architectures for very tough tasks. One-shot NAS is good at finding optimal subnets in Model Space with millions and billions of elements. It is fast, but it is hard to interpret because you get only the best subnet, and you don't know any additional information about another possible solution.

Table 5-5 shows a comparison of different approaches, where

- ✓✓✓: Well suited
- ✓✓: Suited
- ✓: Poorly suited

Table 5-5. *AutoDL approaches comparison*

	HPO	Multi-trial NAS	One-shot NAS
Ease of setup	✓✓✓	✓✓	✓
Search flexibility	✓✓✓	✓✓	✓
Interpretability of results	✓✓✓	✓✓	✓
Designing the neural network search space	✓	✓✓✓	✓✓
Search for small neural networks for simple tasks	✓	✓✓✓	✓
Search for complex architectures for challenging tasks	✓	✓✓	✓✓✓
Optimization of the predesigned architecture for a specific dataset	✓✓✓	✓✓	✓

And here, we again face the problem that there is no unique approach for solving any situation, and the No Free Lunch theorem applies here as well. But understanding how each algorithm acts will help you make the right choice for solving a particular problem.

Summary

One-shot NAS is a very promising area of study. It allows you to find neural network solutions in a reasonable time. Currently, One-shot algorithms can discover completely new architectural keys to solve the most complex problems. This field is developing rapidly and will be a handy tool in any researcher's toolkit. In this chapter, we introduced the basic concepts of One-shot NAS and mastered using two of its algorithms: ENAS and DARTS. This can be a good starting point for putting One-shot NAS into practice. In the next chapter, we will consider the important problem of model compression, which allows you to eliminate unnecessary neural network elements without losing its accuracy.

CHAPTER 6

Model Pruning

Deep learning models have reached significant success in many real-life problems. A lot of devices use neural networks to perform everyday tasks. However, complex neural networks are computationally expensive. And not all devices have GPU processors to run deep learning models. Therefore, it would be helpful to perform model compression methods to reduce the model size and accelerate model performance without losing accuracy significantly. One of the main model compression techniques is model pruning. Pruning optimizes the model by eliminating some model weights. It can eliminate a significant amount of model weights with no negligible damage to model performance. A pruned model is lighter and faster. Pruning is a straightforward approach that can give nice model speedup results.

NNI provides a toolkit to help users to execute model pruning algorithms. `NNI 2.7` version (which is used in this book) supports pruning for the PyTorch framework only. This chapter will study several pruning algorithms and learn how to apply them in practice.

What Is Model Pruning?

Complex neural networks have a lot of layers and many weights. ResNet, DenseNet, and VGGNet take up tens of megabytes to store their weights. Developers often use these architectures to solve many practical problems. But the chosen architecture is often too redundant for the problem being solved, and it contains a lot of extra weights and even extra layers. In practice, you can remove excess weights from the model, which will significantly speed up calculations in the neural network. Sometimes, deep learning models perfectly fit the problem, and they do not have explicit redundancy. Still, they are too computationally heavy to be deployed to the special device they need to act in. In such cases, developers may sacrifice the model accuracy for acceleration.

I. Gridin, *Automated Deep Learning Using Neural Network Intelligence*,
https://doi.org/10.1007/978-1-4842-8149-9_6

Let's describe how the pruning algorithm acts:

1. We take a ready-to-use pre-trained deep learning model we want to prune.

2. Pruning algorithm eliminates (prunes) several weights in the neural network. By eliminating, we mean that the algorithm zeroes these weights. Zeroing the weights significantly speeds up calculations since multiplying by 0 is always 0, and there is no need to do complex calculations. Usually, candidates for elimination are already close to zero, so their zeroing should not seriously affect the model's performance.

3. Pruned model is retrained (fine-tuned) on the same dataset as the original model, but zeroed weights no longer change their values and remain zero all the time. During retraining, active weights try to take over the functions of the removed neurons to continue solving the problem successfully.

Figure 6-1 illustrates the pruning algorithm.

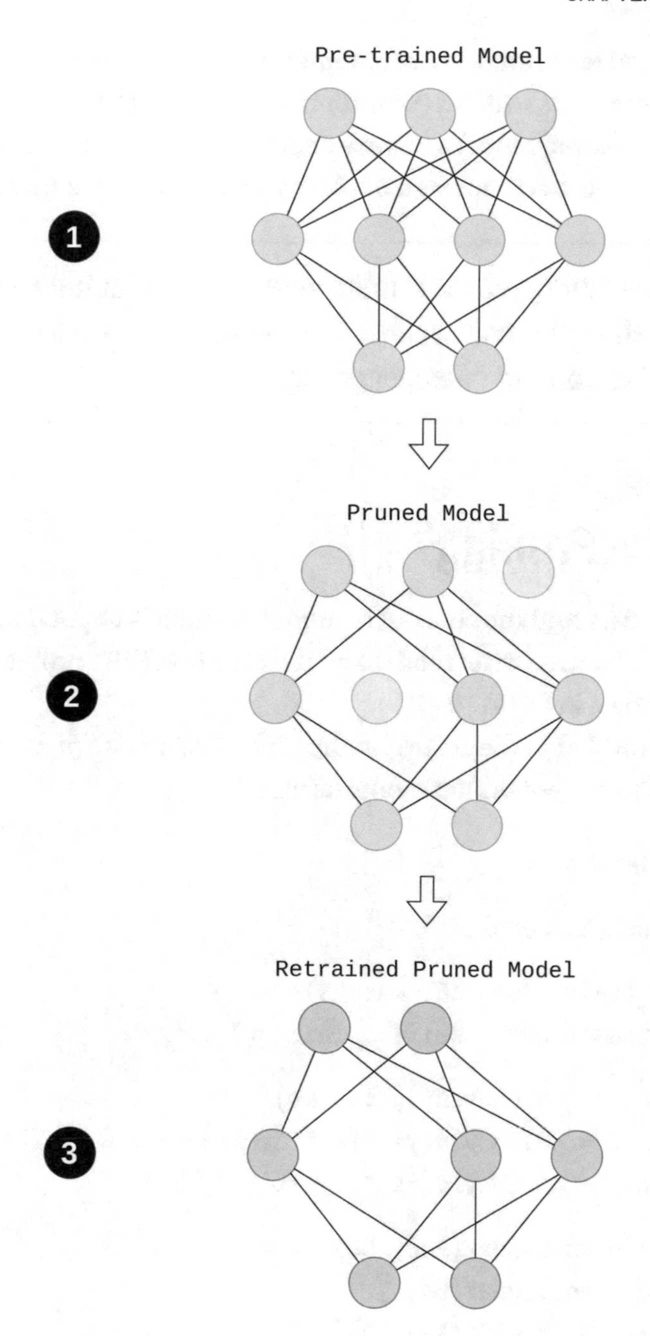

***Figure 6-1.** Pruning algorithm application*

Pruning is a great technique, but sometimes, it degrades the model. It is not always possible to remove weights without compromising the model's accuracy in complex neural networks. However, model accuracy degradation can be minimal, and we'll see further that it is possible to compress a model by 80% with almost no accuracy decrease.

Note Term **fine-tuning** means retraining technique that trains the unpruned weights from their final trained values. In this chapter, we will use the terms **fine-tuning** and **retraining** interchangeably.

LeNet Model Pruning

The best way to understand the concept of model pruning is to put it into practice. Let's get back to the well-known LeNet model and the classic MNIST problem. The LeNet model design is shown in Listing 6-1.

(Full code is provided in the corresponding file: ch6/model/pt_lenet.py.)

`PtLeNetModel` provides familiar model design:

Listing 6-1. LeNet model

```
class PtLeNetModel(nn.Module):

    def __init__(self, fs = 16, ks = 5):
        super(PtLeNetModel, self).__init__()

        self.conv1 = nn.Conv2d(1, fs, ks)
        self.conv2 = nn.Conv2d(fs, fs * 2, ks)
        self.conv3 = nn.Conv2d(fs * 2, fs * 2, ks)

        self.fc1 = nn.Linear(288, 64)
        self.fc2 = nn.Linear(64, 32)
        self.fc3 = nn.Linear(32, 10)

    def forward(self, x):
        x = F.relu(self.conv1(x))
        x = F.relu(self.conv2(x))
        x = F.max_pool2d(x, 2, 2)
```

```
        x = F.relu(self.conv3(x))
        x = F.max_pool2d(x, 2, 2)

        x = torch.flatten(x, start_dim = 1)

        x = torch.relu(self.fc1(x))
        x = torch.relu(self.fc2(x))
        x = torch.relu(self.fc3(x))

        return F.log_softmax(x, dim = 1)
```

Next, we add two helper methods that we will need in the future: `count_total_weights` and `count_total_weights`. `count_nonzero_weights` is a helper method that counts the number of zeros in the neural network's weights, and `count_total_weights` counts the total number of neural network weights.

```
    def count_nonzero_weights(self):
        counter = 0
        for params in list(self.parameters()):
            counter += torch.count_nonzero(params).item()
        return counter

    def count_total_weights(self):
        counter = 0
        for params in list(self.parameters()):
            counter += torch.numel(params)
        return counter
```

The trained `PtLeNetModel` is saved in the following file: ch6/data/lenet.pth. It shows `0.991` accuracy on the test MNIST dataset. Listing 6-2 demonstrates how original pre-trained `PtLeNetModel` can be pruned.

We import necessary modules:

Listing 6-2. LeNet model pruning. ch6/prune/prune.py

```
import os
import random
import matplotlib.pyplot as plt
import torch
from nni.algorithms.compression.v2.pytorch.pruning import LevelPruner
```

```
from ch6.datasets import mnist_dataset
from ch6.model.pt_lenet import PtLeNetModel

CUR_DIR = os.path.dirname(os.path.abspath(__file__))
```

Making script reproducible:

```
# Making script reproducible
random.seed(1)
torch.manual_seed(1)
```

Loading pre-trained `PtLeNetModel`:

```
model = PtLeNetModel()
path = f'{CUR_DIR}/../data/lenet.pth'
model.load_state_dict(torch.load(path))
```

Loading MNIST dataset:

```
train_ds, test_ds = mnist_dataset()
```

Storing accuracy of original model:

```
original_acc = model.test_model(test_ds)
```

Storing nonzero weights of original model:

```
original_nzw = model.count_nonzero_weights()
```

Next, we are pruning original model with one-shot `LevelPruner` (pruning algorithm internals will be explained in the next section):

```
# Pruning Config
prune_config = [{
    'sparsity': .8,
    'op_types': ['default'],
}]
# LevelPruner
pruner = LevelPruner(model, prune_config)
```

Compressing the original model:

```
model_pruned, _ = pruner.compress()
```

Fine-tuning (retraining compressed model):

```
epochs = 10
acc_list = []
for epoch in range(1, epochs + 1):
    model_pruned.train_model(epochs = 1, train_dataset = train_ds)
    acc = model_pruned.test_model(test_dataset = test_ds)
    acc_list.append(acc)
    print(f'Pruned: Epoch {epoch}. Accuracy: {acc}.')
```

Since the compressed model is fine-tuned, let's visualize retraining progress:

```
pruned_nzw = model_pruned.count_nonzero_weights()
plt.title(
    'Fine-tuning\n'
    f'Original Non-zero weights number: {original_nzw}\n'
    f'Pruned Non-zero weights number: {pruned_nzw}')
plt.axhline(y = original_acc, c = "red",
            label = 'Original model accuracy')
plt.plot(acc_list, label = 'Pruned model accuracy')
plt.xlabel('Retraining Epochs')
plt.ylabel('Accuracy')
plt.legend()
plt.show()
```

Visualization result is shown in Figure 6-2.

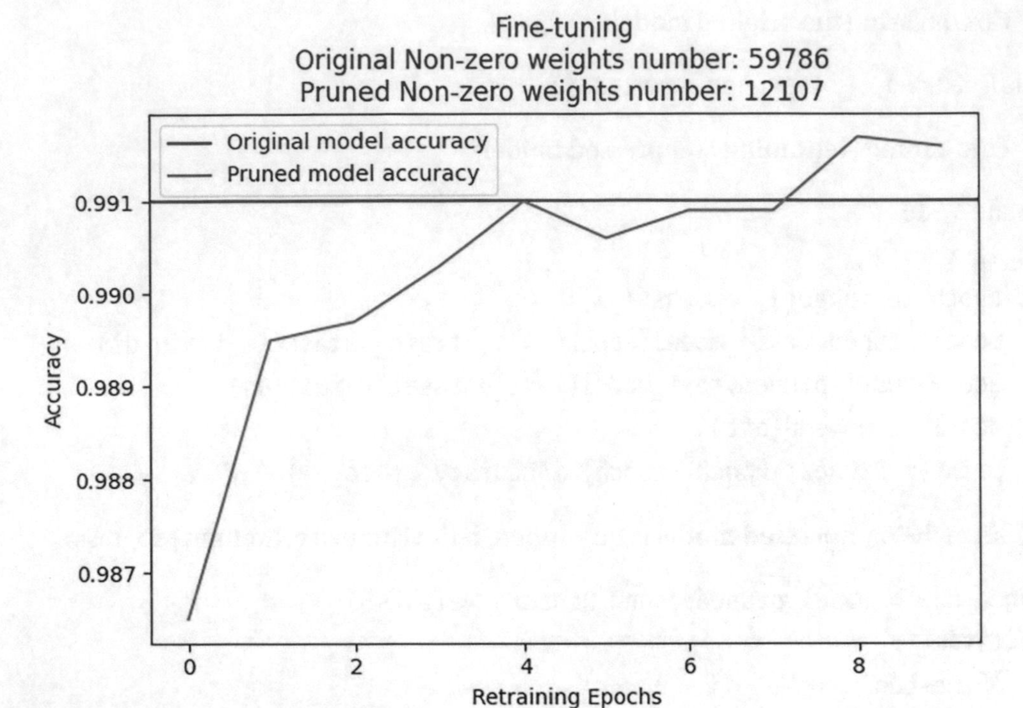

Figure 6-2. *LeNet retraining progress*

Figure 6-2 illustrates fine-tuning of the compressed model. Original trained LeNet model has 59,786 nonzero weights, and compressed LeNetModel has 12,107 nonzero weights only. At the beginning of the retraining, the compressed model is less accurate than the original model. But after 10 training epochs, the compressed model achieves the same accuracy as the original model, having only 12,107 active weights out of 59,780. Obviously, the original LeNet model was redundant and could be replaced by pruned one.

Finally, let's save the pruned model for future usage:

```
model_path = f'{CUR_DIR}/../data/lenet_pruned.pth'
mask_path = f'{CUR_DIR}/../data/mask.pth'
pruner.export_model(
    model_path = model_path,
    mask_path = mask_path
)
```

NNI provides a special wrapper that allows loading pruned models `nni.compression.pytorch.ModelSpeedup` and use them after. In Listing 6-3, we load pruned LeNet model and analyze its characteristics.

(Please install the following package to run this script: `torchsummary`.)

Importing modules:

Listing 6-3. Pruned LeNet model usage. ch6/prune/analyze_pruned.py

```
import os
import torch
from nni.compression.pytorch import ModelSpeedup
from torchsummary import summary
from ch6.model.pt_lenet import PtLeNetModel

CUR_DIR = os.path.dirname(os.path.abspath(__file__))
```

Loading pruned model using `ModelSpeedup` wrapper:

```
dummy_input = torch.randn((500, 1, 28, 28))
model_path = f'{CUR_DIR}/../data/lenet_pruned.pth'
mask_path = f'{CUR_DIR}/../data/mask.pth'
model_pruned = PtLeNetModel()
model_pruned.load_state_dict(torch.load(model_path))
speedup = ModelSpeedup(model_pruned, dummy_input, mask_path)
speedup.speedup_model()
model_pruned.eval()
```

Let's check the accuracy of the pruned model:

```
acc = model_pruned.test_model()
print(acc)
```

The pruned model returns `0.9916` accuracy, the same as the original unpruned model. Also, the pruned model actually shrinks its layers deleting unnecessary weights. The pruned model is less than the original one, and let's examine the difference between them:

```
# Loading Original Model
model_original = PtLeNetModel()
model_path = f'{CUR_DIR}/../data/lenet.pth'
```

```
model_original.load_state_dict(torch.load(model_path))

# Displaying summary of Original and Pruned models
print('==== ORIGINAL MODEL =====')
summary(model_original, (1, 28, 28))
print('========================')

print('====  PRUNED MODEL  =====')
summary(model_pruned, (1, 28, 28))
print('========================')
```

Table 6-1 compares original model and pruned one.

Table 6-1. *Original and pruned model comparison*

Layer	Original Output Shape	Original Size	Pruned Output Shape	Pruned Size
Conv2d-1	[16,24,24]	416	**[14,24,24]**	364
Conv2d-2	[32,20,20]	12,832	[32,20,20]	11,232
Conv2d-3	[32,6,6]	25,632	**[30,6,6]**	24,030
Linear-4	[64]	18,496	**[61]**	16,531
Linear-5	[32]	2,080	**[25]**	1,550
Linear-6	[10]	330	[10]	260

As shown in Table 6-1, pruned model shrinks `Conv2d-1`, `Conv2d-3`, `Linear-4`, and `Linear-5` layers. And this is the primary goal of the pruning algorithm, which eliminates redundancy from the neural network. The earlier example illustrates how we can prune a pre-trained model using NNI. Let's move forward and study pruning algorithms in more detail.

One-Shot Pruners

One-shot pruning algorithms prune weights only once based on a specific metric. Usually, pruned weights are close to zero, and pruner suggests that their removal will not impact the model's accuracy. One-shot pruners act the following way:

- Pruner accepts a model and selects active weights and the weights to be pruned. As a result, the pruner returns a model and a binary mask, where 1 means active weight and 0 means weight to be pruned.
- Original model is speeded up by pruning redundant weights resulting a compressed model.
- Fine-tuning algorithm retrains the compressed model to adjust its weights.

This pruner flow algorithm is depicted in Figure 6-3.

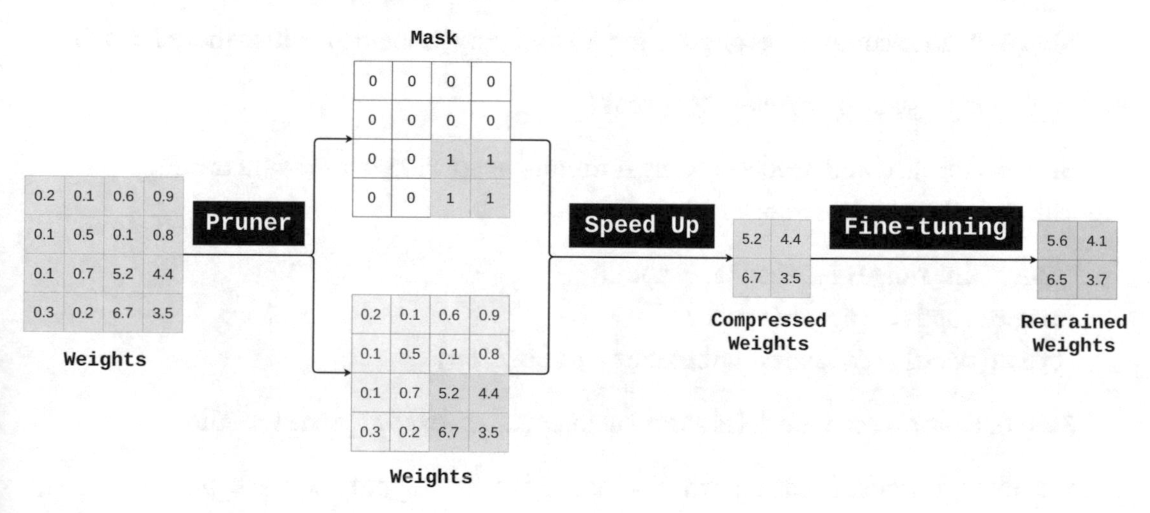

Figure 6-3. *One-shot pruning*

We can define the main steps to implement one-shot pruning algorithms using NNI the following way:

1. Load a pre-trained model
2. Initialize pruner
3. Compress original model
4. Retrain compressed model

5. Save compressed model
6. Load and speed up compressed model

Step 1. Model loading is a native PyTorch operation:

```
model = SomeModel()
model.load_state_dict(torch.load(model_path))
```

Step 2. To initialize the pruner, we must specify the pruner configuration (we will study the pruner configuration in the next section). Here is an example of pruner initialization:

```
prune_config = [{
    'sparsity': .8,
    'op_types': ['default'],
}]
pruner = LevelPruner(model, prune_config)
```

Step 3. Pruner compresses original model applying its logic to reduce model weights:

```
model_pruned, mask = pruner.compress()
```

Step 4. Compressed model is being retrained according to the same training algorithm as the original one:

```
for epoch in range(1, epochs + 1):
    pruner.update_epoch(epoch)
    train(model, dataset, optimizer, criterion)
```

Step 5. Compressed model is stored using `pruner.export_model` method:

```
pruner.export_model(model_path = model_path, mask_path = mask_path)
```

Step 6. Saved model is loaded and wrapped by `nni.compression.pytorch.ModelSpeedup` class:

```
# dummy input tensor
dummy_input = torch.randn((500, 1, 28, 28))
model_pruned = SomeModel()
model_pruned.load_state_dict(torch.load(model_path))
speedup = ModelSpeedup(model_pruned, dummy_input, mask_path)
speedup.speedup_model()
```

And that's it. The compressed model is ready to use! Let's create a helper script that will apply steps 2–6 we defined earlier. Listing 6-4 applies a pruning algorithm and returns a compressed model.

Importing modules:

Listing 6-4. Pruning implementation. ch6/algos/utils.py

```
import copy
import os
import torch
from nni.compression.pytorch import ModelSpeedup
from torchsummary import summary

CUR_DIR = os.path.dirname(os.path.abspath(__file__))
```

`oneshot_prune` method accepts the following parameters:

- `model_original`: Pre-trained original model
- `pruner_cls`: Pruner class
- `pruner_config`: Pruner configuration
- `train_ds`: Train dataset
- `epochs`: Number of training epochs
- `model_input_shape`: Shape of model input

```
def oneshot_prune(
        model_original,
        pruner_cls,
        pruner_config,
        train_ds,
        epochs = 10,
        model_input_shape = (1, 1, 28, 28)
):
    pruner_name = pruner_cls.__name__
    model = copy.deepcopy(model_original)
```

Step 2. Initializing pruner:

```
pruner = pruner_cls(model, pruner_config)
```

Step 3. Compressing original model:

```
model_pruned, mask = pruner.compress()
```

Step 4. Retraining compressed model:

```
for epoch in range(1, epochs + 1):
    model_pruned.train_model(
        epochs = 1,
        train_dataset = train_ds
    )
```

Step 5. Saving compressed model:

```
model_path = f'{CUR_DIR}/../data/{pruner_name}_pruned.pth'
mask_path = f'{CUR_DIR}/../data/{pruner_name}_mask.pth'
pruner.export_model(
    model_path = model_path,
    mask_path = mask_path
)
```

Step 6. Loading and speeding up compressed model:

```
dummy_input = torch.randn(model_input_shape)
model_pruned = model_original.__class__()
model_pruned.load_state_dict(torch.load(model_path))
speedup = ModelSpeedup(model_pruned, dummy_input, mask_path)
speedup.speedup_model()
model_pruned.eval()

return model_pruned
```

Fine. Since we know how to apply one-shot pruning algorithms, let's go further and study some of them.

Pruner Configuration

Each pruner accepts configuration, which specifies its internal logic. Pruner configuration is a `List` of `Dict` entries, and each entry specifies a pruning strategy applied to a specified layer set. Table 6-2 describes pruner configuration parameters.

Table 6-2. *Pruner configuration*

Key	Description
sparsity	Specifies the sparsity for each layer in this configuration entry to be compressed. If `sparsity = 0.8`, then 80% of layer weights will be pruned, and 20% will be left active
op_types	Specifies what types of operations to compress. `'default'` means following the algorithm's default setting. All supported module types for PyTorch are defined package file `nni/compression/pytorch/default_layers.py`: `'Conv1d', 'Conv2d', 'Conv3d', 'ConvTranspose1d', 'ConvTranspose2d', 'ConvTranspose3d', 'Linear', 'Bilinear', 'PReLU', 'Embedding', 'EmbeddingBag'`
op_names	Specifies names of operations to be compressed. If this field is omitted, operations will not be filtered by it
op_partial_names	Operation partial names to be compressed. If `op_partial_names = 'fc_'`, then all layers with the following mask `'fc_*'` will be pruned
exclude	Default is False. If this field is True, it means the operations with specified types and names will be excluded from the compression

Here is an example of the pruning configuration of a one-shot pruner applied to the LeNet model:

```
prune_config = [
    {
        'sparsity': .8,
        'op_types': ['Conv2d'],
    },
```

```
    {
        'sparsity': .6,
        'op_types': ['Linear'],
    },
    {
        'op_names': ['fc3'],
        'exclude': True
    }
]
```

If you don't want to specify a special pruning strategy for each layer type, then you can use the following configuration:

```
prune_config = [
    {
        'sparsity': .8,
        'op_types': ['default']
    }
]
```

Level Pruner

Level pruner is a straightforward one-shot pruner. Sparsity level means prune ratio, that is, `sparsity=0.7` means that 70% of model weight parameters will be pruned. Level pruner sorts the weights in the specified layer by their absolute values. And then mask to zero the smallest magnitude weights until the desired sparsity level is reached.

Level pruner is applied the following way:

```
from nni.algorithms.compression.v2.pytorch.pruning import LevelPruner

prune_config = [{
    'sparsity': .8,
    'op_types': ['default'],
}]

pruner = LevelPruner(model, prune_config)

model_pruned, mask = pruner.compress()
```

Let's apply `LevelPruner` to prune the LeNet model using Listing 6-5.

Importing modules:

Listing 6-5. LevelPruner. ch6/algos/one_shot/level_pruner.py

```
from nni.algorithms.compression.v2.pytorch.pruning import LevelPruner
from ch6.algos.utils import model_comparison, oneshot_prune, visualize_mask
from ch6.datasets import mnist_dataset
from ch6.model.pt_lenet import PtLeNetModel
```

Loading pre-trained model and MNIST dataset:

```
original = PtLeNetModel.load_model()
train_ds, test_ds = mnist_dataset()
```

We will prune `Conv2d` layers with `0.8` sparsity and `Linear` layers with `0.6` sparsity. Also, we will exclude the final classifier linear layer `fc3` from pruning.

```
prune_config = [
    {
        'sparsity': .8,
        'op_types': ['Conv2d'],
    },
    {
        'sparsity': .6,
        'op_types': ['Linear'],
    },
    {
        'op_names': ['fc3'],
        'exclude':  True
    }
]
```

Defining pruner:

```
pruner_cls = LevelPruner
```

Compressing original model using LevelPruner:

```
compressed, mask = oneshot_prune(
    original,
    pruner_cls,
    prune_config,
    train_ds,
    epochs = 10
)
```

Visualizing prune mask:

```
visualize_mask(mask)
```

Figure 6-4 shows the mask of a pruned model. We see that the mask leaves 40% active weights for linear layers and 20% active weights for convolutional layers.

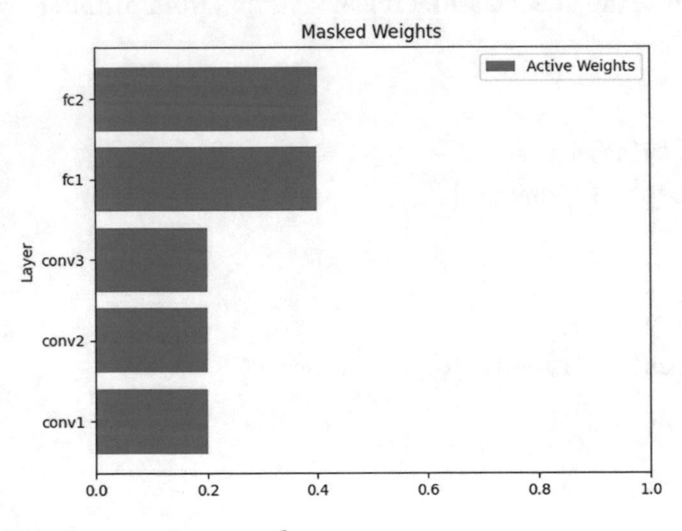

Figure 6-4. *LevelPruner active weights*

Comparing original model and compressed one:

```
model_comparison(original, compressed, test_ds, (1, 28, 28))
```

The original model has `0.991` accuracy, and the compressed one has the same `0.991` accuracy. Table 6-3 compares the architectures of original and compressed models.

Table 6-3. *Original and level pruned model comparison*

Layer	Original Output Shape	Original Size	Pruned Output Shape	Pruned Size
Conv2d-1	[16,24,24]	416	**[14,24,24]**	364
Conv2d-2	[32,20,20]	12,832	[32,20,20]	11,232
Conv2d-3	[32,6,6]	25,632	**[30,6,6]**	24,030
Linear-4	[64]	18,496	[64]	17,344
Linear-5	[32]	2,080	[32]	2,080
Linear-6	[10]	330	[10]	330
Total		59,786		**55,380**

Level pruner compressed the original LeNet model without decreasing its accuracy.

FPGM Pruner

FPGM (Filter Pruning via Geometric Median) Pruner is a one-shot pruner that prunes filters with the smallest geometric median. For more details, please refer to the original paper "Filter Pruning via Geometric Median for Deep Convolutional Neural Networks Acceleration" (`https://arxiv.org/pdf/1811.00250.pdf`).

FPGM Pruner supports `Conv2d`, `Linear` as layers for pruning operation. FPGM Pruner is applied the following way:

```
from nni.algorithms.compression.v2.pytorch.pruning import FPGMPruner

prune_config = [{
    'sparsity': .8,
    'op_types': ['Conv2d'],
}]

pruner = FPGMPruner(model, prune_config)

model_pruned, mask = pruner.compress()
```

Here is an example of FPGMPruner application.

Importing modules:

Listing 6-6. FPGMPruner. ch6/algos/one_shot/fpgm_pruner.py

```
from nni.algorithms.compression.v2.pytorch.pruning import FPGMPruner
from ch6.algos.utils import oneshot_prune, model_comparison, visualize_mask
from ch6.datasets import mnist_dataset
from ch6.model.pt_lenet import PtLeNetModel
```

Loading pre-trained model and MNIST dataset:

```
original = PtLeNetModel.load_model()
train_ds, test_ds = mnist_dataset()
```

Pruning convolutional layers of original model using FPGMPruner with 0.5 sparsity:

```
compressed, mask = oneshot_prune(
    original,
    FPGMPruner,
    [{
        'sparsity': .5,
        'op_types': ['Conv2d'],
    }],
    train_ds
)
```

Visualizing prune mask:

```
visualize_mask(mask)
```

Figure 6-5 shows the mask of the compressed model. We see that the mask leaves 50% active weights for Conv2d layers.

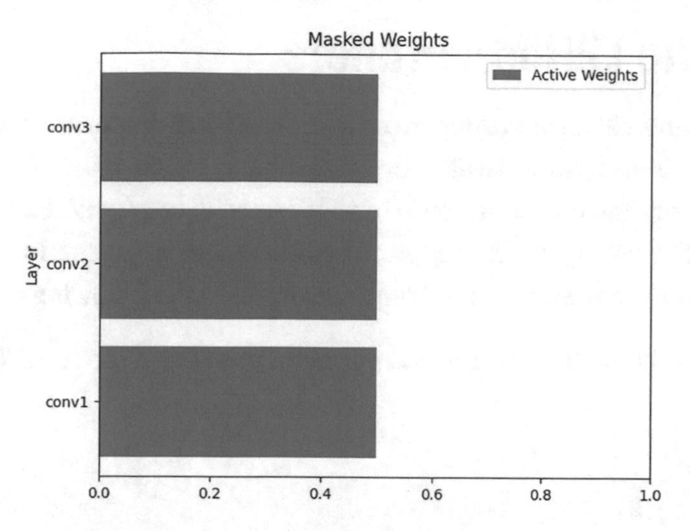

***Figure 6-5.** FPGMPruner active weights*

Comparing original model and compressed one:

```
model_comparison(original, compressed, test_ds, (1, 28, 28))
```

Original model has `0.991` accuracy, while the compressed one has close accuracy `0.9894`. Table 6-4 compares the architectures of original and compressed models.

***Table 6-4.** Original and FPGM pruned model comparison*

Layer	Original Output Shape	Original Size	Pruned Output Shape	Pruned Size
Conv2d-1	[16,24,24]	416	**[8,24,24]**	**208**
Conv2d-2	[32,20,20]	12,832	**[16,20,20]**	**3,216**
Conv2d-3	[32,6,6]	25,632	**[16,6,6]**	**6,416**
Linear-4	[64]	18,496	[64]	9,280
Linear-5	[32]	2,080	[32]	2,080
Linear-6	[10]	330	[10]	330
Total		59,786		**21,530**

Table 6-4 shows that `FPGMPruner` compressed the original model very heavily with almost no loss of accuracy.

L1Norm and L2Norm Pruners

L1Norm Pruner and L2Norm Pruner are one-shot pruners that prune layers according to L_1*-norm* and L_2*-norm*, respectively. For more details, please refer to the original paper "Pruning Filters for Efficient ConvNets" (`https://arxiv.org/pdf/1608.08710.pdf`).

L1Norm and L2Norm pruners support `Conv2d`, `Linear` as a layer for pruning operation. `L1NormPruner` and `L2NormPruner` are applied the following way:

```
from nni.algorithms.compression.v2.pytorch.pruning import L1NormPruner,
L2NormPruner

prune_config = [{
    'sparsity': .8,
    'op_types': ['Conv2d'],
}]

pruner = L1NormPruner(model, prune_config)
# pruner = L2NormPruner(model, prune_config)

model_pruned, mask = pruner.compress()
```

Here is an example of `L2NormPruner` application.

Importing modules:

Listing 6-7. L2NormPruner. ch6/algos/one_shot/l2norm_pruner.py

```
from nni.algorithms.compression.v2.pytorch.pruning import L2NormPruner
from ch6.algos.utils import oneshot_prune, model_comparison, visualize_mask
from ch6.datasets import mnist_dataset
from ch6.model.pt_lenet import PtLeNetModel
```

Loading pre-trained model and MNIST dataset:

```
original = PtLeNetModel.load_model()
train_ds, test_ds = mnist_dataset()
```

Pruning convolutional layers of original model using `L2NormPruner` with `0.7` sparsity:

```
compressed, masks = oneshot_prune(
    original,
    L2NormPruner,
    [{
        'sparsity': .7,
        'op_types': ['Conv2d'],
    }],
    train_ds
)
```

Visualizing prune mask:

```
visualize_mask(masks)
```

Figure 6-6 visualizes the mask of the compressed model. We see that the mask leaves 30% active weights for Conv2d layers.

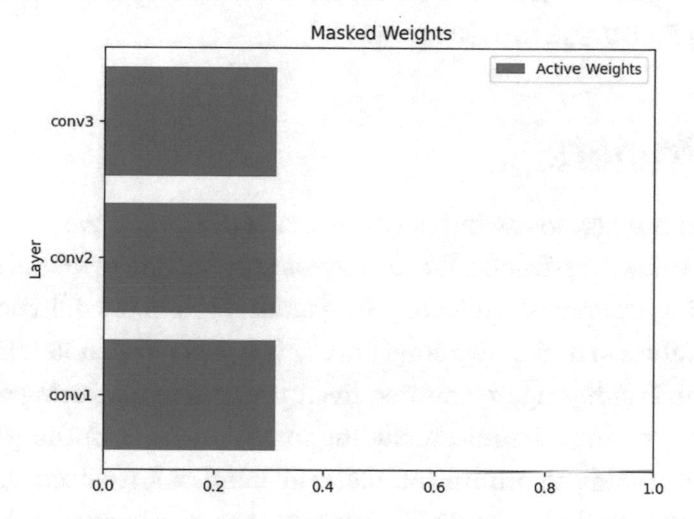

***Figure 6-6.** L2NormPruner active weights*

And let's compare the architectures of the original and compressed model:

```
model_comparison(original, compressed, test_ds, (1, 28, 28))
```

Original model has `0.991` accuracy, while the compressed one degrades to `0.98` accuracy. Table 6-5 compares the architectures of original and compressed models.

Table 6-5. *Original and L2Norm pruned model comparison*

Layer	Original Output Shape	Original Size	Pruned Output Shape	Pruned Size
Conv2d-1	[16,24,24]	416	**[5,24,24]**	**130**
Conv2d-2	[32,20,20]	12,832	**[10,20,20]**	**1,260**
Conv2d-3	[32,6,6]	25,632	**[10,6,6]**	**2,510**
Linear-4	[64]	18,496	[64]	5,824
Linear-5	[32]	2,080	[32]	2,080
Linear-6	[10]	330	[10]	330
Total		59,786		**12,134**

Table 6-5 shows that L2Norm Pruner compressed the original model almost five times, degrading from `0.991` to `0.98` accuracy.

Iterative Pruners

One-shot pruners are easy to use but have one major drawback. We must guess the optimal sparsity values in advance. We usually want to maximize model compression without decreasing accuracy significantly. The natural solution would be to iterate several sparsity values to find the optimal one. This is exactly what iterative pruners were designed for. Pruning algorithms iteratively prune weights during optimization, which control the pruning schedule, including some automatic pruning algorithms. After the iterative pruning algorithm completes all iterations, it selects the best pruned model according to specified score (it is not a necessary accuracy score). Figure 6-7 demonstrates iterative pruning in action.

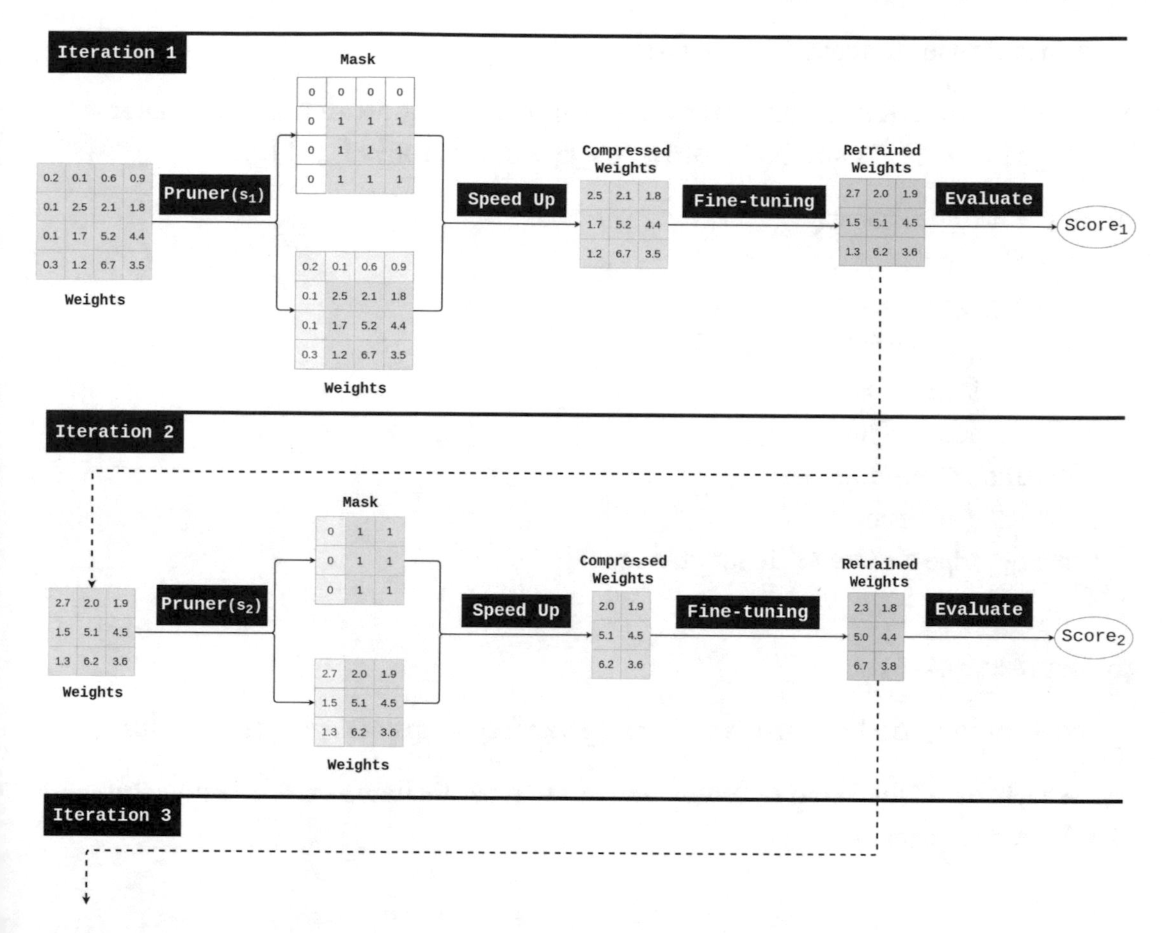

Figure 6-7. *Iterative pruning*

This section will examine two popular iterative tuners: linear pruner and AGP pruner.

Linear Pruner

Linear pruner is an iterative pruner. It will increase sparsity evenly from zero during each iteration. For example, the final sparsity is set as 0.5, and the iteration number is 5, and then the sparsity used in each iteration is [`0.1, 0.2, 0.3, 0.4, 0.5`].

LinearPruner is applied as follows:

```
from nni.algorithms.compression.v2.pytorch.pruning import LinearPruner
config_list = [{ 'sparsity': 0.8, 'op_types': ['Conv2d'] }]

pruner = LinearPruner(
    model = original,
    config_list = config_list,
    pruning_algorithm = 'l1',
    total_iteration = 4,
    finetuner = finetuner,
    evaluator = finetuner,
    speedup = True,
    dummy_input = dummy_input_tensor
)

pruner.compress()

_, compressed, masks, best_acc, best_sparsity = pruner.get_best_result()
```

We will detail the Iterative Tuner configuration parameters in the "Iterative Pruner Configuration" section.

AGP Pruner

AGP is an iterative pruner, in which the sparsity is increased from an initial sparsity value $s_i = 0$ to a final sparsity value s_f over a span of n pruning iterations, starting at training step t_0 and with pruning frequency Δt:

$$s_t = s_f + (s_i - s_f)\left(1 - \frac{t - t_0}{n\Delta t}\right)^3 \quad \text{for} \quad t \in \{t_0,\ t_0 + \Delta t,\ \ldots,\ t_0 + n\Delta t\}$$

For more details, please refer to the original paper "Exploring the efficacy of pruning for model compression" (https://arxiv.org/pdf/1710.01878.pdf).

AGPPruner is applied as follows:

```
from nni.algorithms.compression.v2.pytorch.pruning import AGPPruner
config_list = [{ 'sparsity': 0.8, 'op_types': ['Conv2d'] }]

pruner = AGPPruner(
    model = original,
    config_list = config_list,
    pruning_algorithm = 'l1',
    total_iteration = 4,
    finetuner = finetuner,
    evaluator = finetuner,
    speedup = True,
    dummy_input = dummy_input_tensor
)

pruner.compress()

_, compressed, masks, best_acc, best_sparsity = pruner.get_best_result()
```

We will detail the Iterative Tuner configuration parameters in the "Iterative Pruner Configuration" section.

Iterative Pruner Configuration

Linear pruner and AGP pruner configuration is presented in Table 6-6.

Table 6-6. *Iterative pruner configuration*

Key	Description
`model`	Original PyTorch model
`config_list`	Pruning configuration
`pruning_ algorithm`	(`str`) Specifies pruning algorithm. Supported choices: `level, l1, l2, fpgm, slim, apoz, mean_activation, taylorfo, admm`
`total_ iteration`	(`int`) Total number of iterations
`log_dir`	(`str`) Specifies log directory to save results. You can find the best model `pth` files in this folder
`keep_ intermediate_ result`	(`bool`) If keeping the intermediate result, including intermediate model and masks during each iteration
`finetuner`	(`Optional[Callable[[Module], None]]`) Fine-tuner function that retrains model after each pruning iteration
`speedup`	(`bool`) If set True, speed up the model at the end of each iteration to make the pruned model compact
`dummy_input`	(`Optional[torch.Tensor]`) If `speedup` is True, `dummy_input` is required for tracing the model in speed up
`evaluator`	(`Optional[Callable[[Module], float]]`) Evaluate the pruned model and give a score. If evaluator is None, the best result refers to the latest result
`pruning_params`	(`Dict`) Extra parameters for `pruning_algorithm` implementation

Iterative Pruning Scenarios

This section will study two common use cases for iterative pruning.

Best Accuracy Under Size Threshold Scenario

A common problem is when we need to compress a model not to exceed a specified size threshold. In this case, it is necessary to choose the best model, which will be less than the specified size threshold. Figure 6-8 illustrates how this can be done using iterative pruning.

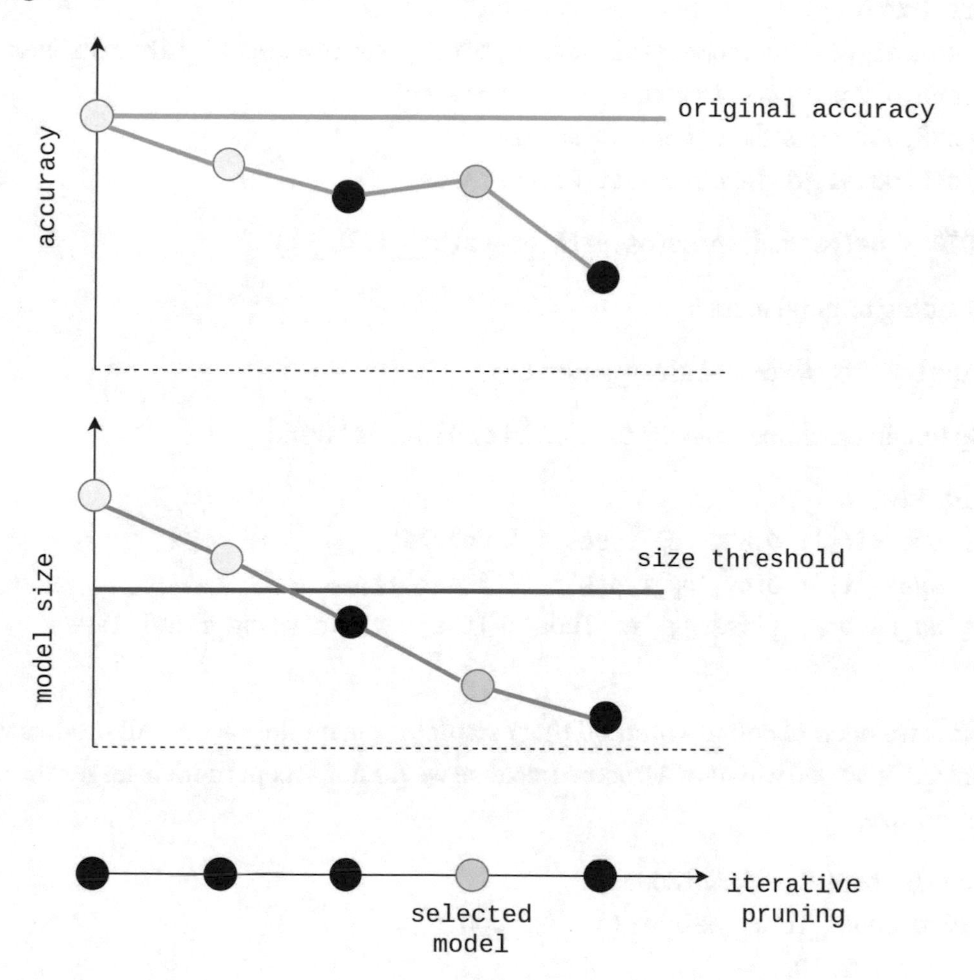

***Figure 6-8.** Best accuracy under size threshold scenario*

Let's handle this scenario to find the best compressed LeNet model that fits 30,000 model size (original size is 59,786) using iterative `LinearPruner`.

Importing modules:

Listing 6-8. Best accuracy under size threshold scenario. ch6/algos/iter/linear_pruner_best_acc_scr.py

```
import os
import torch
from nni.algorithms.compression.v2.pytorch.pruning import LinearPruner
from ch6.algos.utils import model_comparison
from ch6.datasets import mnist_dataset
from ch6.model.pt_lenet import PtLeNetModel

CUR_DIR = os.path.dirname(os.path.abspath(__file__))
```

Loading original model:

```
original = PtLeNetModel.load_model()
```

Specifying maximal sparsity for `Conv2d` and `Linear` layers:

```
config_list = [
    {'sparsity': 0.85, 'op_types': ['Conv2d']},
    {'sparsity': 0.4, 'op_types': ['Linear']},
    {'op_names': ['fc3'], 'exclude': True}  # excluding final layer
]
```

Now we need to define a method that calculates the model's score. All models whose size exceeds 30,000 will have a 0 score because we do not accept models larger than the specified size.

```
def evaluator(m: PtLeNetModel):
    if m.count_total_weights() > 30_000:
        return 0
    return m.test_model()
```

Defining `LinearPruner` with 10 iterations and `L1Norm` pruning algorithm:

```
pruner = LinearPruner(
    model = original,
    config_list = config_list,
```

```
    pruning_algorithm = 'l1',
    total_iteration = 10,
    finetuner = lambda m: m.train_model(epochs = 1),
    evaluator = evaluator,
    speedup = True,
    dummy_input = torch.rand(10, 1, 28, 28),
    log_dir = CUR_DIR  # logging results (model.pth and mask.pth is there)
)
```

Running iterative compression cycle:

```
pruner.compress()
```

Receiving results:

```
_, compressed, masks, best_score, best_sparsity =\
    pruner.get_best_result()
```

The best pruned model is saved in ch6/algos/iter:

```
# Best model saved in CUR_DIR/<date/best_result
print(f'Best Model is saved in: {CUR_DIR}')
```

Now let's analyze results returned by LinearPruner. First, let's display a sparsity of each layer of the best pruned model:

```
print('==========')
print(f'Best accuracy: {best_score}')
print('Best Sparsity:')
for layer in best_sparsity:
    print(f'{layer}')
```

Table 6-7 shows the sparsity of the best pruned model.

***Table 6-7.** Sparsity of the best pruned model for best accuracy under size threshold scenario*

Layer	Sparsity
conv1 (Conv2d)	0.19
conv2 (Conv2d)	0.003
conv3 (Conv2d)	0.16
fc1 (Linear)	0
fc2 (Linear)	0.085

And finally, let's compare the original and the best pruned model:

```
# Displaying comparison
train_ds, test_ds = mnist_dataset()
model_comparison(original, compressed, test_ds, (1, 28, 28))
```

Original model has `0.991` accuracy, while the best compressed one that fits 30,000 size has `0.9913` accuracy. Table 6-8 compares the architectures of original and compressed models.

***Table 6-8.** Original and iteratively pruned by LinearPruner model comparison*

Layer	Original Output Shape	Original Size	Pruned Output Shape	Pruned Size
Conv2d-1	[16,24,24]	416	**[7,24,24]**	**182**
Conv2d-2	[32,20,20]	12,832	**[26,20,20]**	**4,576**
Conv2d-3	[32,6,6]	25,632	**[16,6,6]**	**10,416**
Linear-4	[64]	18,496	**[62]**	**8,990**
Linear-5	[32]	2,080	**[24]**	**1,512**
Linear-6	[10]	330	[10]	**250**
Total		59,786		**25,926**

Great! We have significantly compressed the original LeNet model more than twice without losing accuracy.

Minimal Size Above Accuracy Threshold Scenario

Another case is to compress the model as much as possible while staying above the specified accuracy threshold. Figure 6-9 illustrates how this can be done using iterative pruning.

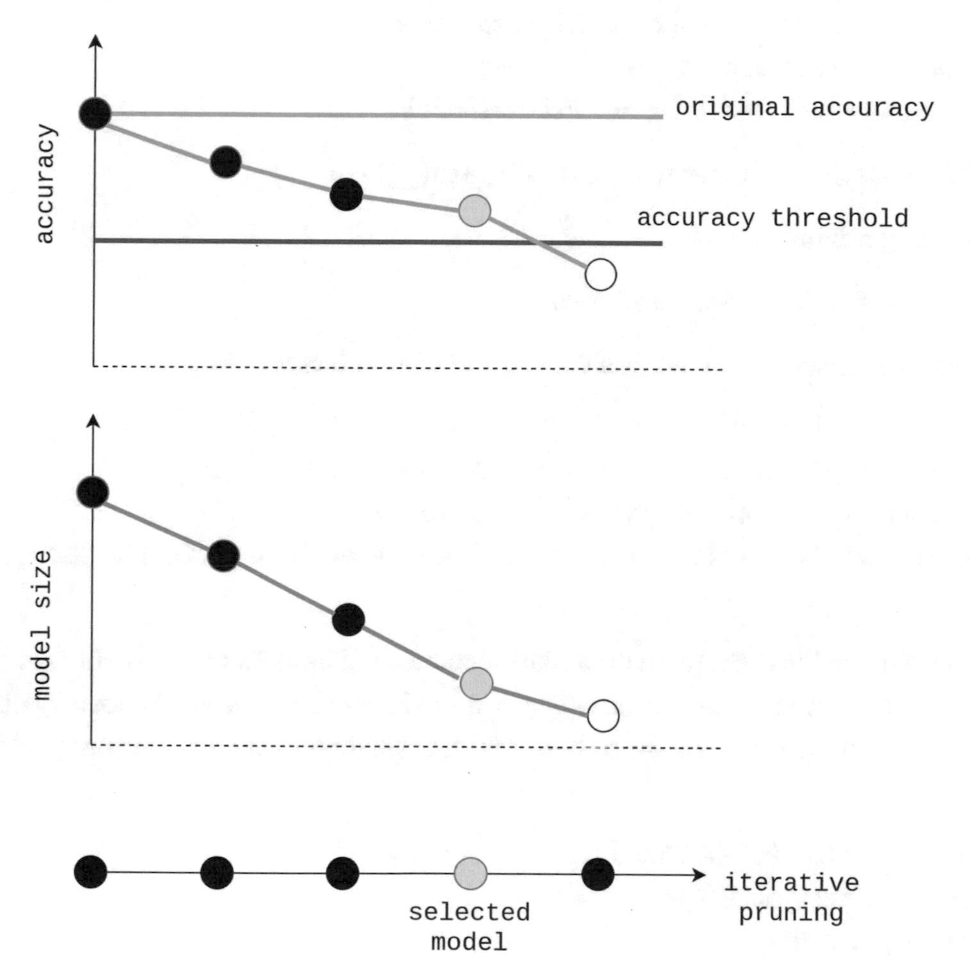

Figure 6-9. *Minimal size above specified accuracy threshold scenario*

Listing 6-9 handles this scenario to find the minimal compressed LeNet model that gives > `0.98` accuracy (original accuracy is `0.991`) using iterative `AGPPruner`.

Importing modules:

Listing 6-9. Minimal size above specified accuracy threshold scenario. ch6/algos/iter/agp_pruner_min_size_scr.py

```
import os
from math import inf
import torch
from nni.algorithms.compression.v2.pytorch.pruning import AGPPruner
from ch6.algos.utils import model_comparison
from ch6.datasets import mnist_dataset
from ch6.model.pt_lenet import PtLeNetModel

CUR_DIR = os.path.dirname(os.path.abspath(__file__))
```

Loading original model:

```
original = PtLeNetModel.load_model()
```

Specifying maximal sparsity for `Conv2d` and `Linear` layers:

```
config_list = [
    {'sparsity': 0.85, 'op_types': ['Conv2d']},
    {'sparsity': 0.4, 'op_types': ['Linear']},
    {'op_names': ['fc3'], 'exclude': True}  # excluding final layer
]
```

Now we need to define a method that calculates the model's score. All models whose accuracy is lower than `0.98` will have a `-inf` score because we do not accept such poor models. If the model performs better than `0.98`, then its score is (`-count_total_weights`):

```
def evaluator(m: PtLeNetModel):
    acc = m.test_model()
    if acc < 0.98:
        return -inf
    return - m.count_total_weights()
```

Defining AGPPruner with 20 iterations and L2Norm pruning algorithm:

```
pruner = AGPPruner(
    model = original,
    config_list = config_list,
    pruning_algorithm = 'l2',
    total_iteration = 20,
    finetuner = lambda m: m.train_model(epochs = 1),
    evaluator = evaluator,
    speedup = True,
    dummy_input = torch.rand(10, 1, 28, 28),
    log_dir = CUR_DIR  # logging results (model.pth and mask.pth is there)
)
```

Running iterative compression cycle:

```
pruner.compress()
```

Receiving results:

```
_, compressed, masks, best_score, best_sparsity =\
    pruner.get_best_result()
```

The best pruned model is saved in ch6/algos/iter:

```
# Best model saved in CUR_DIR/<date/best_result
print(f'Best Model is saved in: {CUR_DIR}')
```

Now let's analyze results returned by AGPPruner. First, let's display a sparsity of each layer of the best pruned model:

```
print('==========')
print(f'Best accuracy: {best_score}')
print('Best Sparsity:')
for layer in best_sparsity:
    print(f'{layer}')
```

Table 6-9 shows the sparsity of the best pruned model.

Table 6-9. *Sparsity of the best pruned model for minimal size above accuracy threshold scenario*

Layer	Sparsity
conv1 (Conv2d)	0.13
conv2 (Conv2d)	0
conv3 (Conv2d)	0.13
fc1 (Linear)	0
fc2 (Linear)	0.03

And finally, let's compare the original and the best pruned model:

```
# Displaying comparison
train_ds, test_ds = mnist_dataset()
model_comparison(original, compressed, test_ds, (1, 28, 28))
```

Original model has `0.991` accuracy and 59,786 size, while the minimal model that exceeds `0.98` accuracy has `0.9883` accuracy and 11,535 size. Table 6-10 compares the architectures of original and compressed models.

Table 6-10. *Original and iteratively pruned by AGPPruner model comparison*

Layer	Original Output Shape	Original Size	Pruned Output Shape	Pruned Size
Conv2d-1	[16,24,24]	416	**[3,24,24]**	**78**
Conv2d-2	[32,20,20]	12,832	**[24,20,20]**	**1,824**
Conv2d-3	[32,6,6]	25,632	**[7,6,6]**	**4,207**
Linear-4	[64]	18,496	**[61]**	**3,904**
Linear-5	[32]	2,080	**[21]**	**1,302**
Linear-6	[10]	330	[10]	**220**
Total		59,786		**11,535**

Yes, the accuracy of the original model degraded from 0.991 to 0.9883, but we compressed the original model more than five times! We made our model lightweight, and now it is more attractive for economical usage.

In this section, we have demonstrated the practical use of iterative pruners in real-world cases. We see that iterative pruning significantly benefits practical deep learning deployment problems.

NNI provides a rich set of pruning algorithms:

- Slim Pruner
- Activation APoZ Rank Pruner
- Activation Mean Rank Pruner
- Taylor FO Weight Pruner
- ADMM Pruner
- Movement Pruner
- Lottery Ticket Pruner
- Simulated Annealing Pruner
- Auto Compress Pruner
- AMC Pruner

Please refer to the official documentation for more details: `https://nni.readthedocs.io`.

Summary

Pruning is an essential part of automated deep learning. It denotes the neural network complexity problem. In addition to the fact that we need an accurate neural network, we also need a lightweight neural network. We will always prefer a neural network with 1M parameters over a neural network with 10M parameters if they have the same accuracy. In this chapter, we have covered the basic principles of model pruning using NNI. Model pruning is a significant direction of neural network optimization that allows the integration of machine learning models into simple devices.

CHAPTER 7

NNI Recipes

In the previous chapters, we studied various NNI features and applications. NNI is a very efficient automated deep learning tool that solves complex deep learning problems. We have witnessed that many NNI experiments can last days or even weeks. Therefore, it is crucial to organize experiments properly. Otherwise, a lot of valuable information and efforts can be lost. On the other hand, NNI uses sophisticated mathematical search algorithms to find the optimal solution in the shortest time in the vast search space. Time is a precious resource. So it is also essential to speed up the NNI execution, which will help maximize the efficiency. It is great to understand the mathematical core of algorithms NNI implements, but it is also important to know how to use NNI effectively.

This chapter will examine patterns and recipes that can help make NNI interactions much more effective. These recipes should help speed up, stabilize, and make research and experiments more developer friendly.

Speed Up Trials

It is essential to speed up the Trial execution in HPO and Multi-trial NAS. The completion of the search algorithm depends on the duration of the Trials, so Trial speed optimization is the first thing a developer should start with. Here, we will mention basic rules a reader should follow to construct a fast Trial.

Use the GPU. One of the most common ways to speed up neural network computations is to use a GPU. Properly configuring the model for GPU usage is the developer's responsibility. If your machine has GPUs, ensure they are utilized during NNI Trial execution.

Do not download dataset twice. A common mistake is downloading heavy datasets without caching them on the disk. Please make sure that the downloaded dataset is cached on disk and the trial does not attempt to download ten gigabytes from the Internet each time it runs a new trial.

I. Gridin, *Automated Deep Learning Using Neural Network Intelligence*,
https://doi.org/10.1007/978-1-4842-8149-9_7

Use the duration panel to determine the longest-running Trials. This can help in finding abnormally long Trials. Figure 7-1 shows NNI duration panel.

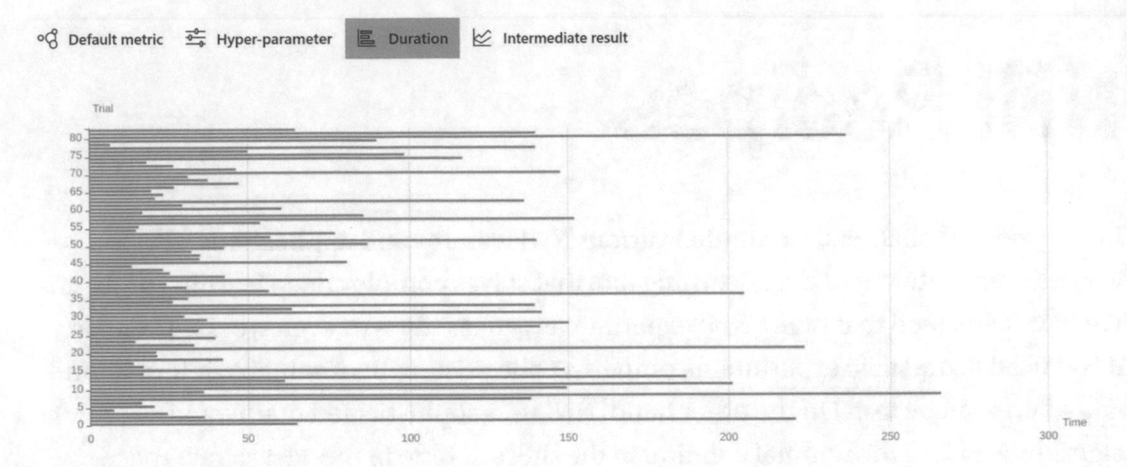

Figure 7-1. *Duration panel*

Use dry trial runs to debug trials. Each trial can be run manually as a Python script, which can help find performance issues and bottlenecks. Try running the Trial several times to check its performance before launching the experiment.

Start–Stop–Resume

Keep in mind that each experiment can be manually stopped and resumed after. All experiment information is stored in the NNI output folder (path is defined by `NNI_OUTPUT_DIR` environment variable, `~/nni-experiments` by default). Therefore, you can stop the experiment at any time using the following command:

```
nnictl stop <experiment_id>
```

and resume it with

```
nnictl resume <experiment_id>
```

You can resume an embedded experiment using the following script:

```
from time import sleep
from nni.experiment import Experiment

experiment = Experiment('local')
```

```
experiment.resume('experiment_id', port)

while True:
    sleep(1)
    if experiment.get_status() == 'DONE':
        break
```

This could be useful when you need to restart an execution machine or an experiment crashed for some reason. It is also possible to resume a finished experiment to re-analyze its results.

Continue Finished Experiment

Suppose the experiment stopped meeting terminal condition (`maxTrialNumber` or `maxExperimentDuration`), but you are not satisfied with the results and want to continue the experiment. In that case, you can resume the finished experiment by changing the terminal condition in WebUI. For example, Figure 7-2 illustrates `maxTrialNumber` update.

Figure 7-2. *Updating maxTrialNumber*

You can also use the WebUI to update the Experiment configuration and search space.

NNI and TensorBoard

NNI can be integrated with TensorBoard. This is very practical if you want to visualize additional Trial metrics. Let's look at an example of integrating NNI with TensorBoard. Make sure tensorboard is installed in your environment. Listing 7-1 illustrates a dummy Trial implementation that writes metrics using TensorBoard format.

Listing 7-1. NNI and TensorBoard. ch7/tb/trial.py

```
import os
from random import random
import nni
```

The simplest way is to use `torch.utils.tensorboard.SummaryWriter` class to export metrics to tensorboard logs:

```
from torch.utils.tensorboard import SummaryWriter
```

Initializing `SummaryWriter`:

```
log_dir = os.path.join(os.environ["NNI_OUTPUT_DIR"], 'tensorboard')
writer = SummaryWriter(log_dir)
```

Trial entry point:

```
if __name__ == '__main__':
    p = nni.get_next_parameter()

    for i in range(100):
```

Calculating dummy metrics:

```
        acc = min((i + random() * 10) / 100, 1)
        loss = max((100 - i + random() * 10) / 100, 0)
```

Writing metrics to tensorboard log:

```
        writer.add_scalar('Accuracy', acc, i)
        writer.add_scalar('Loss', loss, i)

        nni.report_intermediate_result(acc)

    nni.report_final_result(acc)
```

You can run dummy experiment that uses Trial from Listing 7-1 using the following command:

```
nnictl create --config=ch7/tb/config.yml
```

Once the experiment has started, you can go to the Trial jobs panel on Trails detail page, select Trials you want to analyze, and click TensorBoard button, as shown in Figure 7-3.

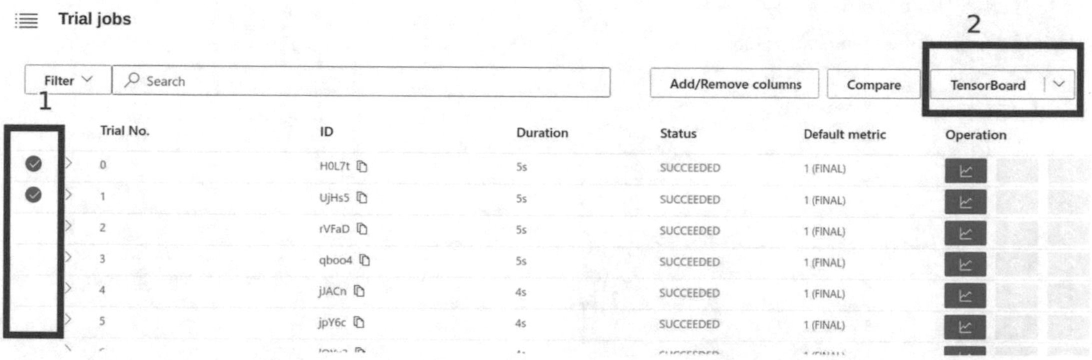

Figure 7-3. *Launching TensorBoard*

After clicking the TensorBoard button, NNI starts the TensorBoard process passing Trial log directories as its input and redirects the browser to its web page. Figure 7-4 shows TensorBoard panel with metrics we have collected during dummy trials we defined in Listing 7-1.

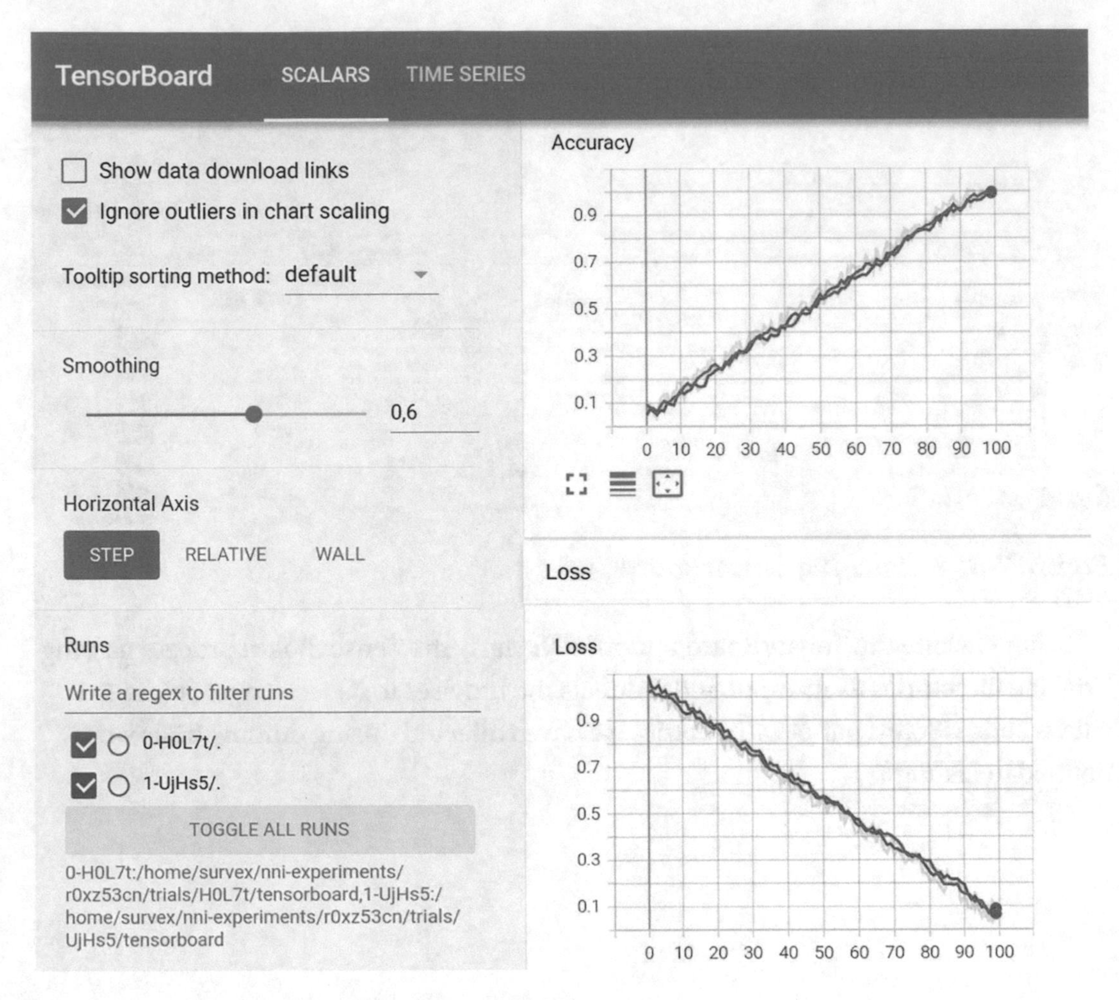

***Figure 7-4.** TensorBoard displaying Trial metrics*

NNI runs an actual TensorBoard process, so you can stop it when you're done with it, as shown in Figure 7-5.

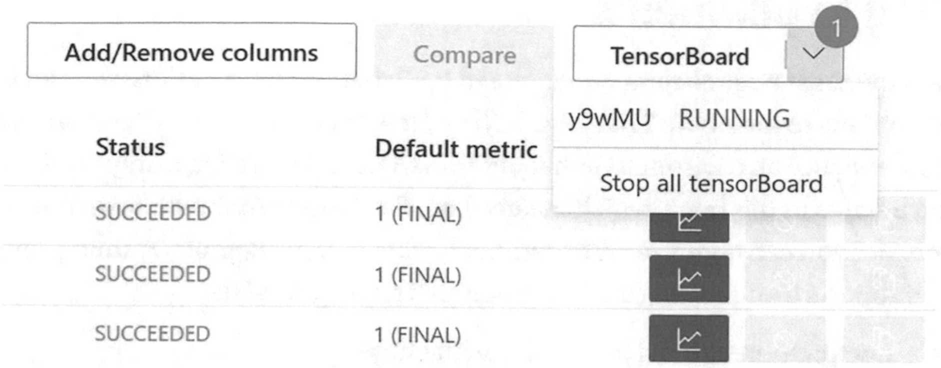

Figure 7-5. Stopping TensorBoard process

Integrating TensorBoard in your NNI Experiments can help you analyze Trial results and whole Experiment progress.

Move Experiment to Another Server

All information about the Experiment is stored in the `NNI_OUTPUT_DIR/<experiment_id>` folder (`~/nni-experiments/<experiment_id>` by default), so you can easily move the experiment data to another server to resume it there. You just need to stop the Experiment, move the folder to another server, and resume the Experiment. Figure 7-6 illustrates this approach.

`Server A: nnictl stop <experiment_id>`

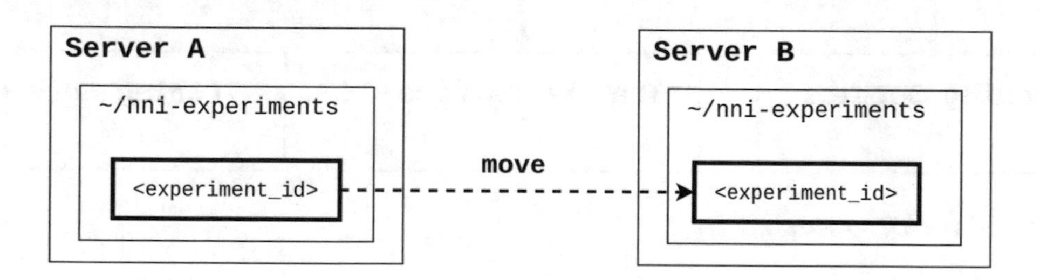

`Server B: nnictl resume <experiment_id>`

***Figure 7-6.** Moving Experiment to another server*

You can use this trick if you want to move your experiment to a more powerful server or if you want to share the results of an experiment.

Scaling Experiments

Scaling is the most natural approach to speed up Experiment execution. You can use multiple servers to distribute Trial jobs. NNI implements the Training Service concept. Training Service is an environment that performs Trial jobs. We have only used the Local Training Service in this book, which means that all calculations are done on the local machine. But you can organize an Experiment using various Remote Training Services. NNI 2.7 supports the following environments as Training Services:

- **Local**: Running Trial jobs on local machine
- **Remote**: Running Trial jobs on remote machine using ssh
- **OpenPAI**: Running Trial jobs on Microsoft Open Platform for AI server
- **AML**: Running Trial jobs on Azure Machine Learning server
- **Hybrid**: Allows setting several different Training Services

Many search algorithms allow concurrent Trial execution, so you can horizontally scale the experiment, significantly increasing its speed. Figure 7-7 illustrates this concept.

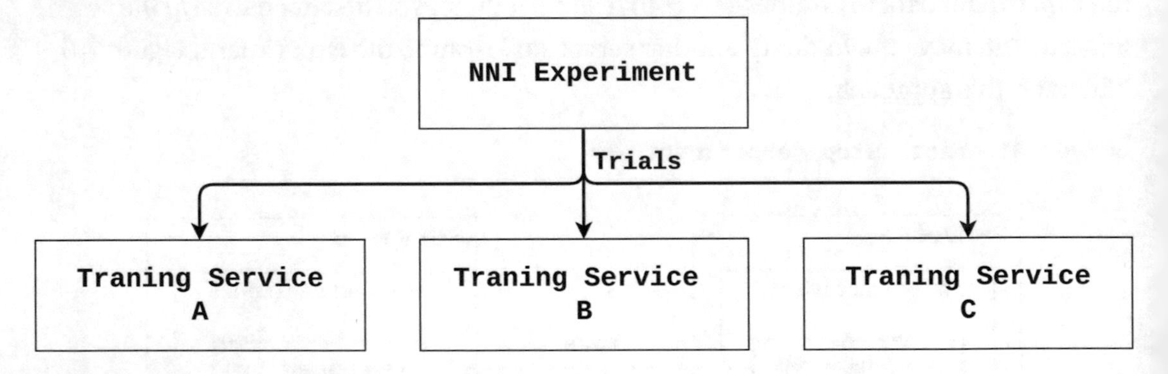

Figure 7-7. *Horizontal scaling*

Let's look at an example configuration that uses the Remote Training Service.

Common configuration part:

```
trialConcurrency: 4
maxTrialNumber: 100
searchSpace:
```

```
    x:
      _type: quniform
      _value: [1, 100, 0.1]

trialCodeDirectory: .
trialCommand: python3 trial.py
tuner:
  name: Random
```

Remote Training Service settings. `nniManagerIp` is used as the host Experiment address to send metrics from Trial jobs running on remote machines:

```
nniManagerIp: <nni_host_ip> # example: 10.10.120.20

trainingService:
  platform: remote
```

Listing remote machines with ssh access:

```
  machineList:

    - host: <remote1_ip> # example: 10.10.120.21
      user: <remote1_ssh_user> # example: nni_user
      password: <remote1_ssh_pass> # example: nni_user_pass
      pythonPath: <remote1_ssh_pass> # example: /opt/python3/bin

    - host: <remote2_ip> # example: 10.10.120.22
      user: <remote2_ssh_user> # example: nni_user
      password: <remote2_ssh_pass> # example: nni_user_pass
      pythonPath: <remote2_ssh_pass> # example: /opt/python3/bin
```

You can apply Remote Training Service in embedded (stand-alone) NNI mode as follows:

```
# Loading Packages
from nni.experiment import Experiment, RemoteConfig, RemoteMachineConfig
from pathlib import Path
```

Remote Training Service parameters:

```
nni_host_ip = '10.10.120.20'
remote_ip = '10.10.120.21'
remote_ssh_user = 'nni_user'
remote_ssh_pass = 'nni_pass'
remote_python_path = '/opt/python3/bin'
```

Common Experiment configuration:

```
# Defining Search Space
search_space = {
    "x": {"_type": "quniform", "_value": [1, 100, .1]}
}

# Experiment Configuration
experiment = Experiment('remote')
experiment.config.experiment_name = 'Remote Experiment'
experiment.config.trial_concurrency = 4
experiment.config.trial_command = 'python3 trial.py'
experiment.config.trial_code_directory = Path(__file__).parent
experiment.config.max_trial_number = 1000
experiment.config.search_space = search_space
experiment.config.tuner.name = 'Random'
```

Remote Training Service configuration:

```
experiment.config.nni_manager_ip = nni_host_ip
remote_service = RemoteConfig()
remote_machine = RemoteMachineConfig()
remote_machine.host = remote_ip
remote_machine.user = remote_ssh_user
remote_machine.password = remote_ssh_pass
remote_machine.python_path = remote_python_path
remote_service.machine_list = [remote_machine]
experiment.config.training_service = remote_service
```

Starting NNI:

```
http_port = 8080
experiment.start(http_port)
```

Processing event loop:

```
while True:
    if experiment.get_status() == 'DONE':
        break
```

The remote server must have the same Python environment installed as the Experiment host server. NNI copies the experiment information to the remote server and executes the Trial jobs during the experiment. Here is an example of a Trial process executed on a remote server:

```
python3 -m nni.tools.trial_tool.trial_runner --job_pid_file /tmp/nni-
experiments/5jixfy3o/envs/XfJ9j/pid
```

NNI provides rich explanations concerning Training Services. Please refer to the official documentation for more details: `https://nni.readthedocs.io/`.

Shared Storage

NNI scaling we considered in the previous section has one serious drawback. Training Services return only Trial metrics (`nni.report_intermediate_result` and `nni.report_final_result`) to Experiment host server. All Trial logs are stored on the machine they are executed as shown in Figure 7-8.

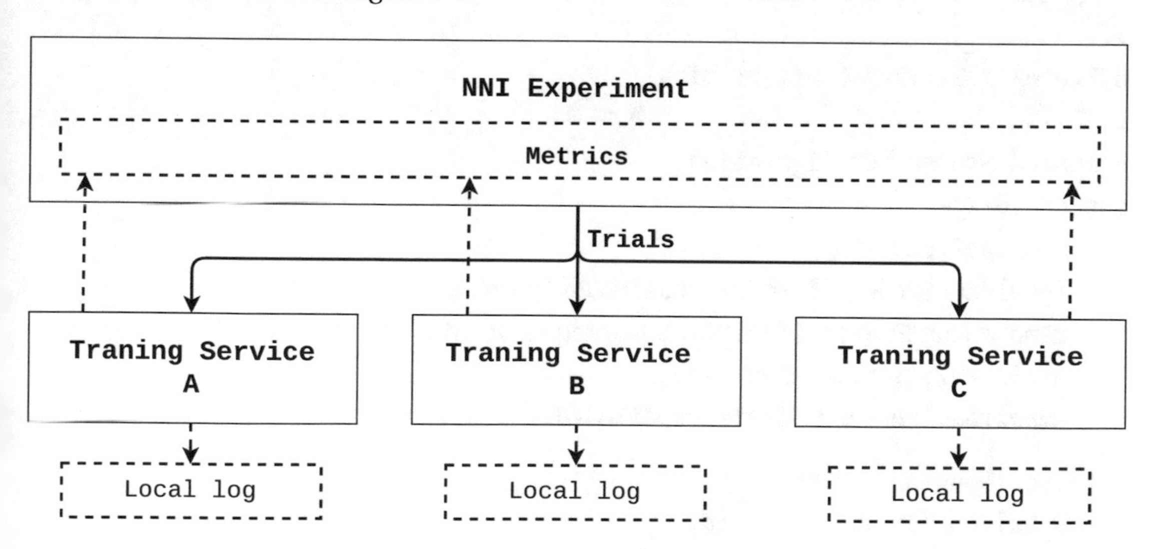

Figure 7-8. *Local logging*

This is not convenient because the logs are located in different places.

To solve this problem, NNI provides a Shared Storage implementation that allows you to store all Trial logs in one place, accessible to the NNI Experiment. Figure 7-9 depicts architecture of NNI Experiment with Shared Storage.

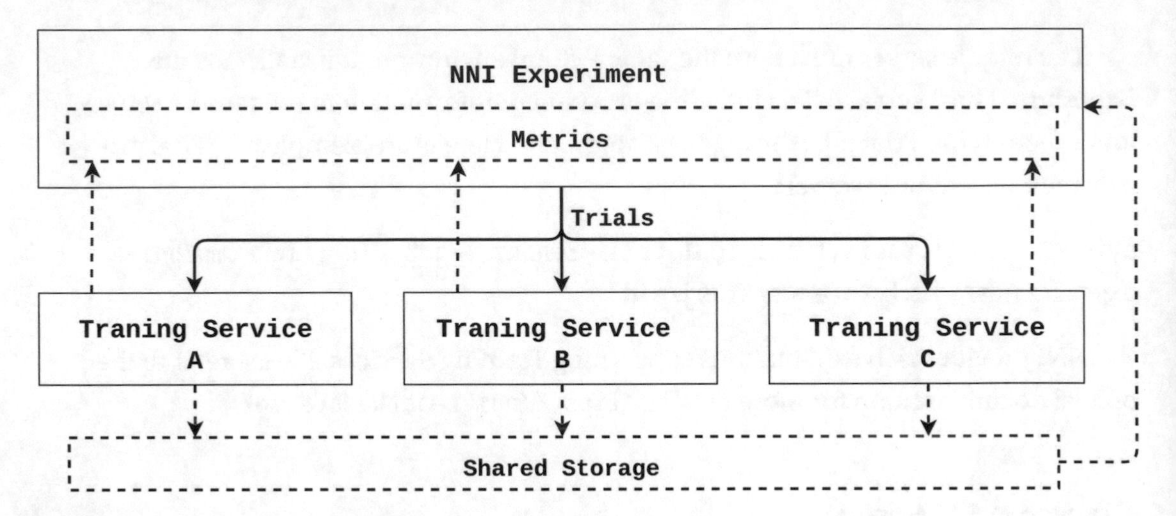

Figure 7-9. *Shared Storage*

There are two ways to implement Shared Storage in Experiment: NFS and Azure Blob. Here is a sample configuration for NFS Shared Storage:

```
# Experiment Configuration
...
# Training Service Configuration
...
# Shared Storage Configuration
sharedStorage:
    storageType: NFS
    localMountPoint: ${your/local/mount/point}
    remoteMountPoint: ${your/remote/mount/point}
    nfsServer: ${nfs-server-ip}
    exportedDirectory: ${nfs/exported/directory}

    localMounted: nnimount
    # Values for localMounted:
```

```
# usermount: means you have already mount this storage on
  localMountPoint
# nnimount: means nni will try to mount this storage on localMountPoint
# nomount: means storage will not mount in local machine, will support
  partial storages in the future
```

Please refer to the official documentation for more details concerning Shared Storage implementation: `https://nni.readthedocs.io/`.

One-Shot NAS with Checkpoints and TensorBoard

All of the patterns we have studied in this chapter are suitable for HPO and Multi-trial NAS, but they are useless for One-shot NAS. Indeed, One-shot NAS is an exceptional case, and it has the following limitations:

- Cannot be stopped and resumed
- Does not have a visualization of the training process
- Cannot be restored after an error occurs
- Cannot be transferred to another server

These are pretty serious limitations, making working with such an effective method much more difficult. Let's try to eliminate all these limitations on the example of PyTorch LeNet Supernet (ch7/one_shot_nas/pt_lenet.py) and `DartsTrainer` implementation.

`DartsTrainer` accepts a user-defined method that calculates Supernet accuracy during the training process. To visualize the training process, we can use tensorboard logging with `SummaryWriter` that logs each training iteration's accuracy. Listing 7-2 demonstrates how to visualize training progress.

(Full code is provided in the corresponding file: ch7/one_shot_nas/pt_utils.py.)

Listing 7-2. Injecting TensorBoard logging in accuracy method

```
from torch.utils.tensorboard import SummaryWriter
```

Initializing `SummaryWriter`:

```
cd = os.path.dirname(os.path.abspath(__file__))
dt = datetime.now().strftime("%Y-%m-%d_%H-%M-%S")
```

```
tb_summary = SummaryWriter(f'{cd}/runs/{dt}')
iter_counter = 0
```

Method that calculates `Supernet` accuracy for `DartsTrainer`:

```
def accuracy(output, target, topk = (1,)):

    global iter_counter
...
```

Calculating `results`:

```
    res = dict()
    for k in topk:
        correct_k = correct[:k].reshape(-1).float().sum(0)
        accuracy = correct_k.mul_(1.0 / batch_size).item()
```

Passing accuracy to TensorBoard logs:

```
        tb_summary.add_scalar('darts_lenet', accuracy, iter_counter)
        iter_counter += 1

        res["acc{}".format(k)] = accuracy

    return res
```

Injecting TensorBoard logging in accuracy method allows visualizing Supernet training progress. But the main problem is that the One-shot NAS process is very fragile. If the server crashes or an `OutOfMemory` error occurs, the One-shot NAS process will be stopped without the possibility of resuming. This is a very serious risk of losing valuable results and time. Let's try to solve this problem. We need to mention that One-shot NAS training uses a training loop like most neural networks. The One-shot NAS algorithm takes specific actions to converge to the optimal subnet during each training epoch. When we run the trainer `fit` method, we run the training loop. But this training loop can be run several times, and each time a new training loop will train the Supernet model from `DartsTrainer`. So we can split one fit method with `num_epochs` = 50 into five fit methods with `num_epochs` = 10. Figure 7-10 illustrates this concept.

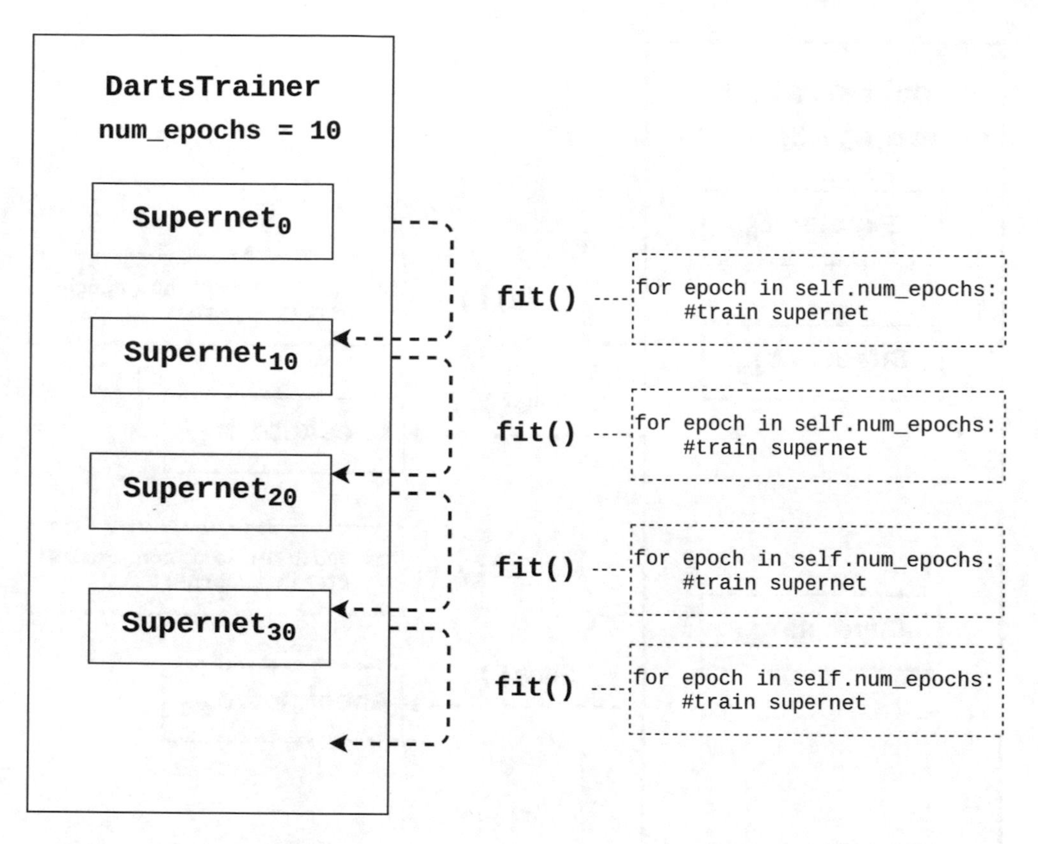

Figure 7-10. *Shared Storage*

And what we can do is dump a binary image of `DartsTrainer` between training subcycles creating checkpoints, as shown in Figure 7-11.

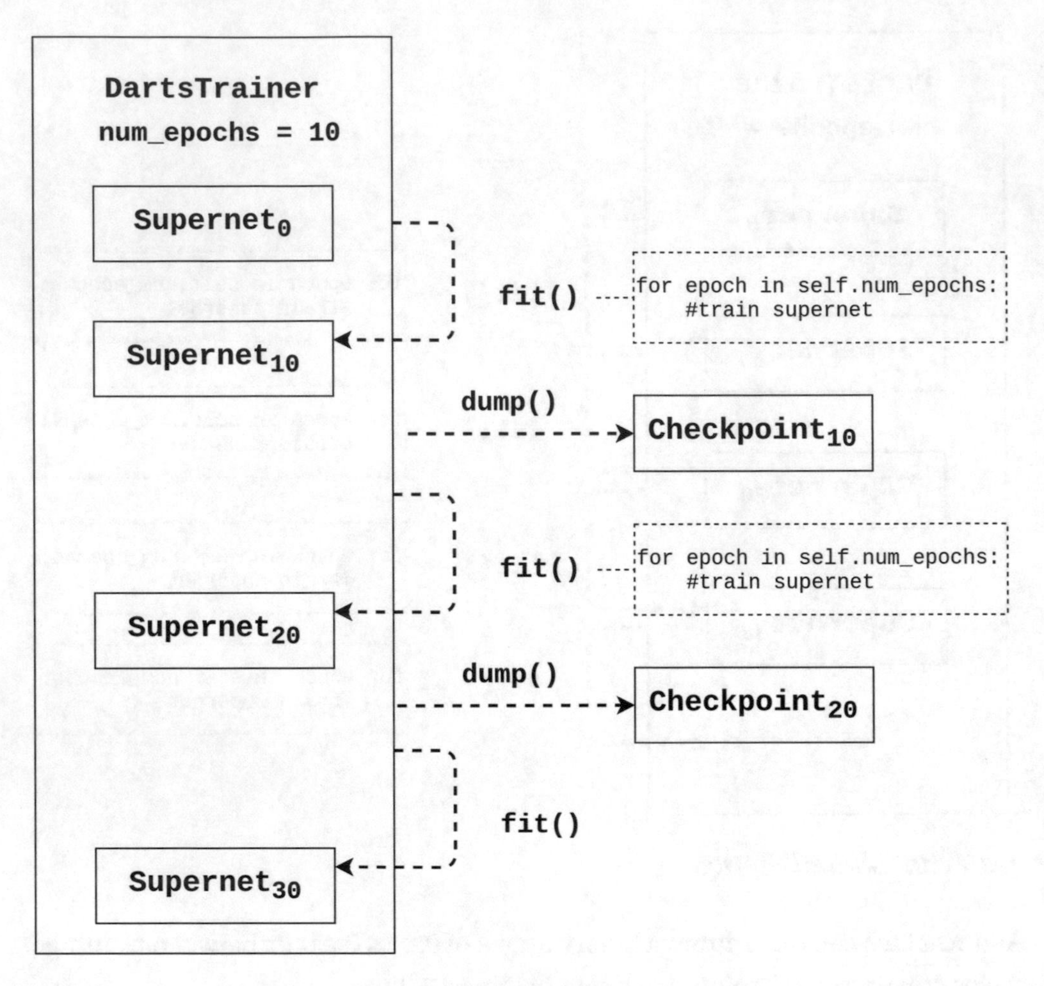

Figure 7-11. *One-shot NAS with checkpoints*

This trick allows us to solve all main problems concerning the One-shot NAS process:

- Process can be resumed if machine crashes.
- Process can be stopped manually and resumed after.
- Checkpoint can be moved to another machine and resumed there.

Listing 7-3 demonstrates how One-shot NAS process with checkpoint dumping can be implemented.

We are using `pickle` to dump trainer binary image (please install this package if necessary):

Listing 7-3. One-shot NAS with checkpoints. ch7/one_shot_nas/darts_train_with_checkpoint.py

```
import pickle
```

Importing other modules:

```
import os
from os.path import exists
import torch
import torch.nn as nn
import ch7.datasets as datasets
from nni.retiarii.oneshot.pytorch import DartsTrainer
from ch7.one_shot_nas.pt_lenet import PtLeNetSupernet
from ch7.one_shot_nas.pt_utils import accuracy
```

Specifying trainer checkpoint path:

```
cd = os.path.dirname(os.path.abspath(__file__))
trainer_checkpoint_path = f'{cd}/darts_trainer_checkpoint.bin'
```

Following method created DartsTrainer for LeNet Supernet:

```
def get_darts_trainer():
    # Supernet
    model = PtLeNetSupernet()

    # Dataset
    dataset_train, dataset_valid = datasets.get_dataset("mnist")

    # Loss Function
    criterion = nn.CrossEntropyLoss()

    # Optimizer
    optim = torch.optim.SGD(
        model.parameters(), 0.025,
        momentum = 0.9, weight_decay = 3.0E-4
    )

    # Trainer params
    num_epochs = 0
```

```
    batch_size = 256
    metrics = accuracy

    # DARTS Trainer
    darts_trainer = DartsTrainer(
        model = model,
        loss = criterion,
        metrics = metrics,
        optimizer = optim,
        num_epochs = num_epochs,
        dataset = dataset_train,
        batch_size = batch_size,
        log_frequency = 10,
        unrolled = False
    )

    return darts_trainer
```

The following method trains Supernet for specified number of epochs and dumps trainer:

```
def train_and_dump(darts_trainer, epochs):
    """
    Trains Supernet according to DARTS algorithm
    """
    darts_trainer.num_epochs = epochs
    darts_trainer.fit()

    with open(trainer_checkpoint_path, 'wb') as f:
        pickle.dump(darts_trainer, f)

    return darts_trainer
```

And here is the main script that loads the trainer from binary checkpoint if necessary and splits the whole training loop into multiple subcycles:

```
if __name__ == '__main__':

    if exists(trainer_checkpoint_path):
```

```
        with open(trainer_checkpoint_path, 'rb') as f:
            trainer = pickle.load(f)
    else:
        trainer = get_darts_trainer()

    for _ in range(10):
        trainer = train_and_dump(trainer, epochs = 5)

    print(f'Best model: {trainer.export()}')
```

Now let's run the script (ch7/one_shot_nas/darts_train_with_checkpoint.py) we examined in Listing 7-3. We can visualize One-shot NAS process with TensorBoard:

```
tensorboard --logdir=ch7/one_shot_nas/runs/
```

And now, we can monitor Supernet training progress using the following link: `http://localhost:6006/#scalars`. Figure 7-12 demonstrates TensorBoard web page.

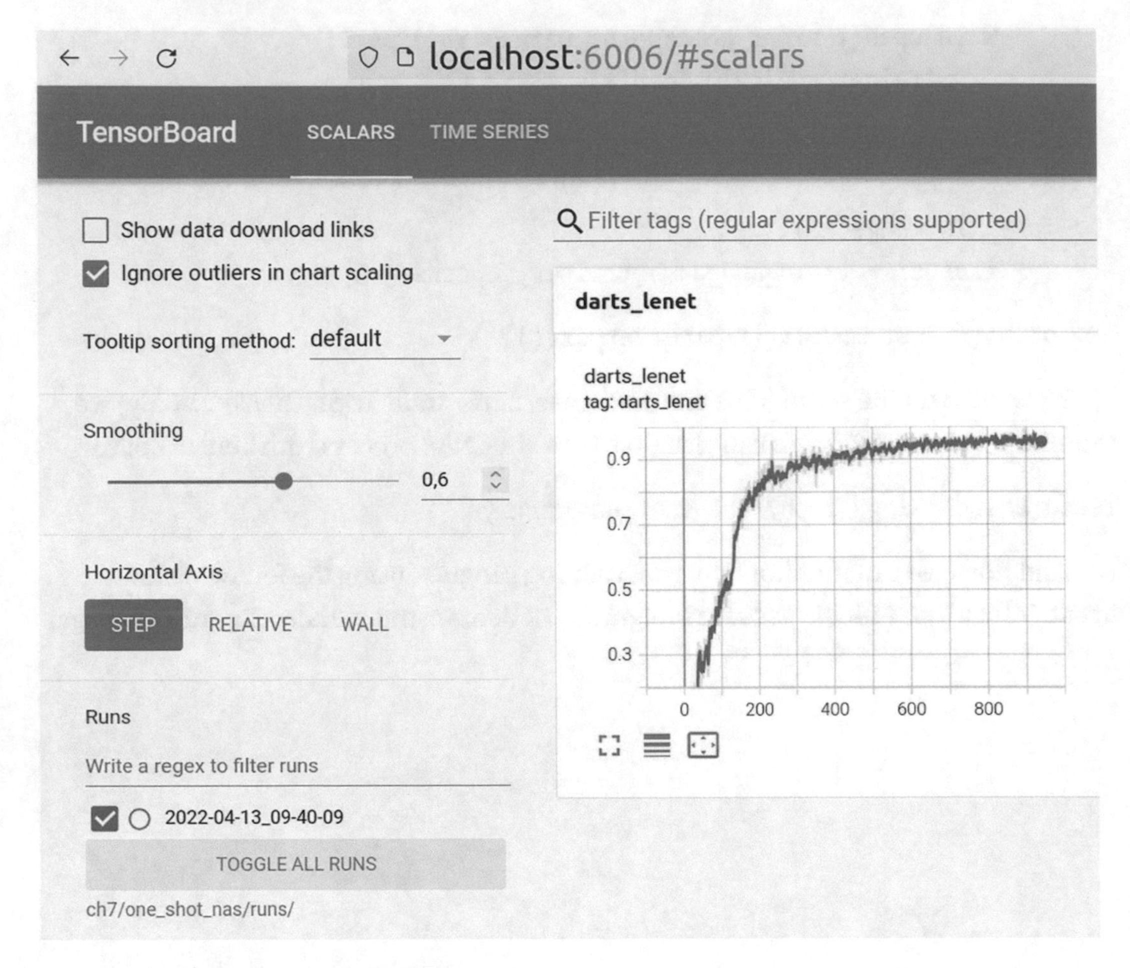

***Figure 7-12.** Supernet training progress*

But the most important thing is that you can stop the execution of ch7/one_shot_nas/darts_train_with_checkpoint.py script and then run it again. DartsTrainer will be restored from the binary checkpoint file ch7/one_shot_nas/darts_trainer_checkpoint.bin and continue Supernet training. Figure 7-13 shows that `DartsTrainer` continues training from the checkpoint, not from scratch.

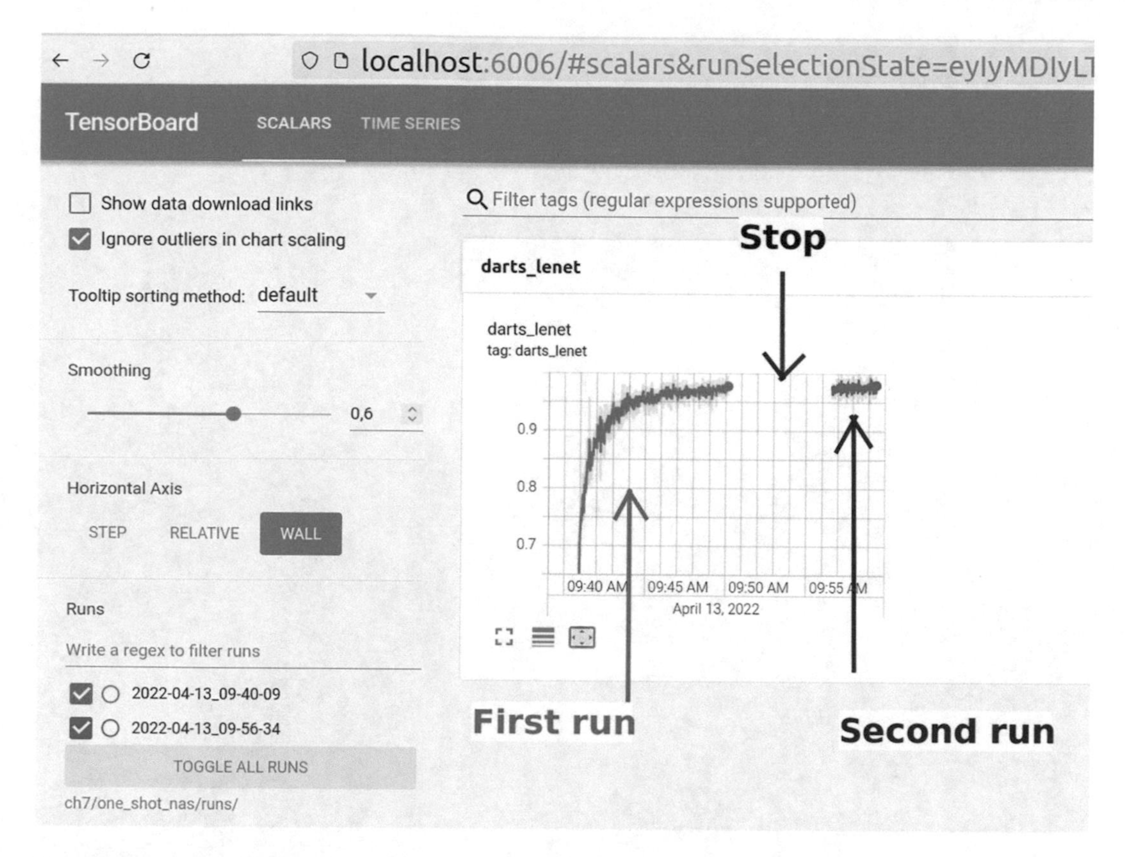

Figure 7-13. *One-shot NAS training resume*

This technique allows the implementation of really long-term One-shot NAS Experiments without fear that the Experiment or the server will crash. You can also stop and resume Experiment at any convenient time.

Summary

This chapter examined several tricks and patterns that can facilitate your user experience. NNI is an open source developer-friendly framework, so you can implement your own ideas and approaches in your research and experiments. In this chapter, we complete the book. I hope that you now can appreciate the effectiveness of using the NNI framework, and it can become an indispensable tool for your daily research activity.

Index

I. Gridin, *Automated Deep Learning Using Neural Network Intelligence*,
https://doi.org/10.1007/978-1-4842-8149-9

E

F

G

H

I, J, K

L

O

P

Q

R

S

T

U, V

W, X, Y, Z

Printed in the United States
by Baker & Taylor Publisher Services